Memoirs of the American Mathematical Society
Number 132

Hewitt Kenyon and A. P. Morse

Web Derivatives

Published by the
American Mathematical Society
Providence, Rhode Island
1973

ABSTRACT

After some measure-theoretic preliminaries, the notion of *web* is defined and used to construct a dimension- and metric-free setting for the differentiation of non-negative valued set functions. Conditions which guarantee the frequent existence of the derivative of one such function with respect to another are examined in detail, and then strengthened to permit integration of the derivative in case the denominator function is a measure. A Lebesgue decomposition of a function into a part which is the integral of its derivative and a remainder with derivative almost everywhere zero is obtained. The resulting theory is applied to differentiation of indefinite integrals, complex valued functions of bounded variation on the reals, and others.

WEB DERIVATIVES

HEWITT KENYON AND A. P. MORSE

CONTENTS

1. INTRODUCTION.

In this paper we have constructed a dimension- and metric-free theory of differentiation of set functions which simplifies and generalizes much existing theory, including that of blankets. Our mechanism for set convergence is a *web*, described in 3.2.1 in the language of *Runs*[1]. The casual reader may read "filter" or "net" in place of "run", but anyone interested in details must refer to our short paper, *Runs*, which we wrote as a preliminary to the present paper.

The strands of our argument in the highly technical Sections 2–7 are isolated and minutely examined in a way which, we are well aware, obscures, at first, the pattern of the whole cloth. We shall purposely here in the introduction reconstruct concepts at some length in less technical language in hope the reader can better see the weave of the fabric.

A *run*, R, is such a nonvacuous relation that corresponding to each x and y in its domain there is a z in its domain for which

$$(x,t) \in R \text{ and } (y,t) \in R$$

Received by the editors June 21, 1971.

whenever t is such that $(z,t) \in R$. That is to say, t is past both x and y whenever t is past z. Runs are direct generalizations of directions, but differ from the latter in that the domain of a run might not intersect its range.

If R is a run, then the family of sets of the form

$$\beta = \text{the set of points } y \text{ for which } (x,y) \in R,$$

where x is in the domain of R, is a filter base which does the work of R, unless, as sometimes happens, each set β is too large to be a member of a family. For example, R might be the set of pairs (B, C) for which C is finite and $B \subset C$; here R is the run used to define unordered summation.

Each web W is a function which associates with every point x of its domain A a run $W(x)$ of sets which can be thought of as converging in some abstract sense to x, and which will usually, but not necessarily, be related to x and A. The family of all sets thus converging toward the various points of A is the *reach* of W. Specifically, $\beta \in$ reach W if and only if, for some x in A, β is in the range of $W(x)$.

In much the same way that subblankets are formed from blankets, our theory requires the frequent formation of *cowebs* of a given web by restricting it in domain or reach. To explain and exploit cowebs it is convenient to have a simple device for restricting a relation in range. Thus we agree that strn R B is the set of pairs in R of the form (x,y) where x is arbitrary and y is in B. With a web, W, and a family, F, we associate a special web, cwb W F, which is such a function V that:

the domain of V consists of those x in the domain of W for which $W(x)$ is frequently in F;

for each x in the domain of V,

$$V(x) = \text{strn } W(x)\ F.$$

We feel it noteworthy that the restriction in range of $W(x)$ is across the board and does not vary with x. Webs like V are special cowebs of W from which all other cowebs descend hereditarily. Thus W' is a coweb of W if and only W is a web and for some family F,

$$W' \subset \text{cwb } W\ F;$$

accordingly cwb W F is an extension of W'.

If φ is a measure, then the *realm* of φ is the union of the sets in its domain. Since for us all measures are outer measures, the domain of φ then consists of all subsets of its realm. The range of φ is contained in $[0, \infty]$.

To help describe the various classes of functions which will be differentiated with respect to each other we let the *domain'* of a function f consist of those β in the domain of f for which $|f(\beta)| < \infty$. We now agree that

$$\text{tile } \varphi \; W$$

consists of those non-negative valued functions f for which:

> φ is a measure and W is a web;
>
> some countable subfamily of the domain of f covers φ almost all of the domain of W;
>
> for φ almost all x in the domain of W, $W(x)$ is eventually in both the domain' of f and the domain of φ;
>
> corresponding to each $\epsilon > 0$, each β in the domain of f, and each coweb W' of W having its domain included in β, there is a countable, not necessarily disjointed, subfamily G of the reach of W' which covers φ almost all of the domain of W' in such a way that

$$0 \leqslant \Sigma \; \alpha \in G \; f(\alpha) \leqslant f(\beta) + \epsilon .$$

It turns out that if tile φ W is nonvacuous, then the domain of W is included in the realm of φ.

The above description of tile φ W is not a mere translation of 4.3.2 into less technical language. It is instead an equivalent recasting of 4.3.2 in the light of 4.5.1 and 4.13. Other informal descriptions will follow this pattern.

In the theorem below and later in the introduction, $\overline{h}$ is such a function on the domain of φ that for each β in the domain of φ, $\overline{h}(\beta)$ is the infimum of numbers of the form

$$\Sigma \; \alpha \in G \; h(\alpha),$$

where G is a countable subfamily of the domain of h covering φ almost all of β.

THEOREM. If

$$g \in \text{tile}\ \varphi\ W, \quad h \in \text{tile}\ \varphi\ W,$$

then for $\overline{h}$ almost all x in the domain of W the limit as β runs along $W(x)$ of the ratio

$$\frac{g(\beta)}{h(\beta)}$$

is a finite non-negative number.

The above theorem is a more easily understood consequence of 5.3. The derivative whose frequent existence it assures will be in the future

$$\mathsf{D}\ W\ g\ h\ x.$$

We propose now to explain 6.100. In the theorem below the measure $\overline{g}$ is such a function on all subsets of the real finite numbers, rf, that for $A \subset \text{rf}$, $\overline{g}(A)$ is the infimum of numbers of the form

$$\Sigma\ \beta \in G\ V(\beta),$$

where G is a countable subfamily of K covering all but a countable subset of A. Also in the theorem below

$$\mathsf{D}\ f\ /\ g\ x = \lim_{t \to 0} \frac{f(x+t) - f(x)}{g(x+t) - g(x)}.$$

THEOREM. If K is the family of ordinary closed intervals of finite positive length, f and g are complex valued functions of bounded variation on each member of K, w is the variation function of g, V is such a function on K that

$$V(\beta) = w(b) - w(a)$$

whenever $-\infty < a < b < \infty$ and $\beta = [a,b]$, then for $\overline{g}$ almost all real finite numbers x:

.1 $$|\mathrm{D}\, f/w\, x\,| < \infty;$$

.2 $$|\mathrm{D}\, g/w\, x\,| = 1;$$

.3 $$|\mathrm{D}\, f/g\, x\,| < \infty.$$

Conclusion .3 answers affirmatively a question we have been asked in the past. Quite possibly it is in the literature but we have not found it.

Looking forward to the integration of the derivative, we agree that

$$\text{goodtile } W$$

consists of such measures φ that:

$$\varphi \in \text{tile } \varphi\; W;$$

each set in the domain of φ is included in a φ measurable set with the same φ measure; for φ almost all x in the domain of W, $W(x)$ is eventually φ measurable. That the above description accords with 6.11.2 is a consequence of 6.10.1, 6.17 and 6.18. Here in 6.18 assume f is φ and M is the family of φ measurable sets.

We agree

$$\text{solidtile } W$$

consists of such φ in goodtile W that the domain of W includes φ almost all of the realm of φ.

If

$$\varphi \in \text{solidtile } W, \quad h \in \text{tile } \varphi\; W,$$

then: because of 7.4

$$\mathrm{D}\; W\; h\; \varphi\; x \text{ is } \varphi \text{ measurable in } x\,;$$

because of 7.6 every φ measurable set is also $\overline{h}$ measurable; because of 7.9

$$\overline{h}(B) = \int B; (\mathrm{D}\ W\ h\ \varphi\ x)\ \varphi\ \mathrm{d}x$$

whenever B is a φ measurable set.

Listed in Section 4 of *Product Measures*(2) are the not unusual properties we assume for our integral. Among these is the assumption that if the integral exists as a finite or infinite number, then the integrand is zero except on a countable union of measurable sets of finite measure.

We now examine the differentiation of the integral. We first agree that

goodpave W

consists of those φ in goodtile W for which for φ almost all x in the domain of W, $W(x)$ is eventually in the family of such sets B that corresponding to each coweb W' of W having its domain included in B, there is a countable, disjointed subfamily G of the reach of W' which covers φ almost all of the domain of W'.

That this description of goodpave W accords with 6.11.3 is a consequence of 6.21, 6.13.7 and 6.9.

If $\varphi \in$ goodtile W and u is a non-negative valued function for which

$$0 \leqslant \int u(x)\ \varphi\ \mathrm{d}x \leqslant \infty,$$

and g is the function on the family of all φ measurable sets for which

$$g(\beta) = \int \beta; u(x)\ \varphi\ \mathrm{d}x$$

whenever $\beta \in$ dmn g, then:

if $g \in$ tile φ W then, according to 7.21,

$$\mathrm{D}\ W\ g\ \varphi\ x = u(x)$$

for φ almost all x in the domain of W;

if u is a bounded function then, according to 6.39 and 6.37,

$$g \in \text{tile } \varphi\, W;$$

if

$$\varphi \in \text{goodpave } W \quad \text{and} \quad \int u(x)\, \varphi\, \mathrm{d}x < \infty$$

then, according to 6.40 and 6.38,

$$g \in \text{tile } \varphi\, W.$$

In order to describe helpfully 6.100 and some of the results in Section 7 as quickly as we could, we have bypassed a very useful family, subtile φ W, which not only involves the interesting family undersum M, but has a much stronger flavor of eventuality than tile φ W. We now rectify the omission.

The family undersum M consists of such non-negative valued functions f that:

if the empty set 0 is in the domain of f, then $f(0) = 0$;

corresponding to each $\epsilon > 0$, each B in the domain of f, each countable subfamily C of M, there is a countable subfamily K of M for which

K is a refinement of C,

K covers exactly as much as C,

$f(B) + \epsilon \geqslant \Sigma\ \beta \in K\ f(B \cap \beta) \geqslant 0.$

The family subtiled f φ W consists of such sets B in the domain of f that:

f is a non-negative valued function;

φ is a measure; W is a web;

corresponding to each $\epsilon > 0$, each coweb W' of W having domain included in B, each set β belonging to the domain of f and including the domain of W', there is a countable subfamily G of the reach of W' which covers φ almost all the domain of W' in such a way that

$$\Sigma\ \alpha \in G\ f(\alpha) \leqslant f(\beta) + \epsilon.$$

Finally,

$$\text{subtile } \varphi\ W$$

consists of such functions f that there is a family M for which:

$f \in$ undersum M;

some countable subfamily of subtiled $f\ \varphi\ W$ covers φ almost all of the domain of W;

for φ almost all x in the domain of W, $W(x)$ is eventually in the domain$'$ of f, the domain of φ, subtiled $f\ \varphi\ W$, and M.

That the above description of subtile φ W accords with 6.7 is a consequence of 4.13 and 6.13.4.

From 6.17 we learn

$$\text{subtile } \varphi\ W \subset \text{tile } \varphi\ W;$$

from 6.18 we learn conversely that if

$$f \in \text{tile } \varphi\ W$$

and there is a family M for which

$$f \in \text{undersum } M$$

and, for φ almost all x in the domain of W,

$$W(x) \text{ is eventually in } M,$$

then

$$f \in \text{subtile } \varphi\, W.$$

Because of the elementary novelty of measure-theoretic 2.8 and because of the crucial use of 2.8 in the proof of 7.3, we now propose to make 2.8 more accessible.

THEOREM. If φ and f are measures with the same domain, every set in the domain of f is included in a φ measurable set with the same f measure, every set of φ measure zero is of f measure zero, and

$$A \subset A', \quad \varphi(A) = \varphi(A') < \infty, \quad T \text{ is } \varphi \text{ measurable},$$

then

$$f(T \cap A) = f(T \cap A').$$

We think it interesting that in 2.8 no mention is made of an f measurable set.

Although we reject any rigid classification of variables, it does happen that a notational rhythm pervades most of the paper. With a few exceptions it turns out that:

φ and ψ are measures;
S is the realm of φ;
f, g, h are akin to measures;
W, W', V, V' are webs;
$A, A', B, T, \alpha, \beta, \gamma$ are in the domain of φ and often also in the domain of f and the reach of W;
x is in the domain of W and the realm of φ;
F, F', G, G', H, H' are subfamilies of the domain of φ and quite possibly of the reach of W;

in connection with 2.2.12, $\overline{f} = \overline{\mathrm{m}}\, f\, \varphi$;
u is a candidate for integration.

Countercurrently K, C, M and N are wild cards.

In Section 2 we establish the measure-theoretic foundation and terminology for our theory.

In connection with 2.1.2 and 2.2.4 we hope to help some readers by pointing out that, in our eyes,

$$\sigma G\ \beta = \sigma G \cap \beta = \beta \cap \sigma G = \beta\ \sigma G.$$

In Section 3 we define and establish the basic properties of webs and their derivatives, assuming only that functions differentiated with respect to each other are non-negative.

To facilitate the study of the effect of various stronger hypotheses on the functions to be differentiated, we define in Section 4 a good many families of such functions with special properties which we study separately and in several combinations. The most important of these is tile φ W, which is described above. The measure φ is used at first only for the purpose of specifying the domain of φ and negligible sets of zero measure.

In Section 5 we examine the fundamentals of differentiation. Theorem 5.22 at the end of the section shows that a suitable function may be decomposed in the manner of Lebesgue into a part which turns out in 7.11 to be the integral of its derivative and a part with zero derivative almost everywhere.

In order to apply our results of Section 5, we must of course develop techniques for discovering when functions in which we are interested belong to tile φ W and related families. Continuing to maintain a rather general point of view, we study this problem from many angles in Section 6.

After formulating and studying some new families of functions, we examine in 6.24 through 6.62 a theory inspired by indefinite integration and specifically applied thereto in Theorems 6.39, 6.40 and 6.59. We are indebted to S. Kakutani for suggesting a problem which he had already solved directly but felt might be amenable to solution by our methods; our treatment of this problem led to the results in 6.63 through 6.83. In 6.84 through 6.100 we take advantage of the work of D. C. Peterson(3) to find applications to interval functions on the line and in the plane. Noteworthy here is our application of the fundamental Theorem 6.96 to Theorem 6.100, mentioned above, in which we differentiate one function of bounded variation with respect to another without ever using the countability of their discontinuities. Some readers may welcome the omissions suggested by the paragraph which begins Section 6.

In Section 7, Theorem 7.4, which guarantees the measurability of the derivative, and Theorem 7.6 are the key results. The proof of Lemma 7.3 long eluded us. We have greatly simplified an earlier proof by the discovery and application of Theorem 2.8. In our approach to Theorem 7.6 we were greatly helped by Trevor McMinn's Theorem 2.16.

In the course of our paper we have used over 150 special notational constants to condense some fairly complex notions into wieldy form. These enable us to state all theorems completely without implicitly assumed hypotheses, and to write fairly detailed proofs without adding unduly to the total length of the article. In order to cope with the resulting style, most readers will need to make frequent use of the definitional index we have provided in Section 8.

Because of the rather self-contained nature of our results, our list of references, which appears in Section 9, is not extensive. A comprehensive review of the field and a large bibliography are provided in *Differentiation of Integrals* by A. M. Bruckner in Part II of the November 1971 issue of the American Mathematical Monthly.

A defect in type face is illustrated by the

CONJECTURE. $\nu = v$.

Since it takes a sharp eye to spot the invalidity of the above conjecture we shall somewhat artificially avoid technical use of its *first* symbol which, unfortunately, is lower case italic vee.

We are grateful for support given by the Miller Institute, the University of California, and the George Washington University.

2. MEASURES. Many of the theorems of this section are either well known or easily checked, and are stated without proof.

2.1 DEFINITIONS.

.1 $\sim A$ = complement $A = \mathsf{E}\, x\ (x \notin A)$ = the set of points x such that x is not a member of A.

.2 $\sigma F = \mathsf{E}\, x\ (x \in \beta$ for some $\beta \in F)$.

.3 $\pi F = \mathsf{E}\, x\ (x \in \beta$ for every $\beta \in F)$.

.4 sb A = subset $A = \mathsf{E}\, B\ (B \subset A)$ = the family of sets B such that $B \subset A$.

.5 sp A = superset $A = \mathsf{E}\, B\ (B \supset A)$.

.6 sng A = singleton $A = \mathsf{E}\, B\ (B = A)$.

.7 rlm $\varphi = \sigma$ dmn φ = the union of the domain of φ.

.8 dmn$'$ $\varphi = \mathsf{E}\, A \in$ dmn φ $(|\varphi(A)| < \infty)$.

.9 Join$''$ F = the family of sets of the form σH, where H is a countable subfamily of F.

.10 Meet$''$ F = the family of sets of the form $\sigma F \cap \pi H$, where H is a countable subfamily of F.

.11 meet$'$ F = the family of sets of the form πH, where H is a finite nonempty subfamily of F.

.12 ω = the set of non-negative integers.

.13 ω' = the set of integers.

.14 zr $f = \mathsf{E}\, x\ (f(x) = 0)$.

Since the empty set 0 is countable, $\sigma 0 = 0$, and $\pi 0$ = the universe, we infer from .9 and .10 above that $0 \in$ Join$''$ F and $\sigma F \in$ Meet$''$ F for every family F. On the other hand, σF is not a member of meet$'$ F unless $\sigma F \in F$.

2.2 DEFINITIONS.

.1 gauge = $\mathsf{E}\, g$ (g is such a function that

$$0 \leqslant g(\beta) \leqslant \infty$$

whenever $\beta \in$ dmn g).

.2 Msr $S = \mathsf{E}\, \varphi \in$ gauge (dmn φ = sb S, and

$$\varphi(A) \leqslant \Sigma \beta \in F\ \varphi(\beta)$$

whenever F is a countable family for which

$$A \subset \sigma F \subset S).$$

.3 Measure = $\mathsf{E}\ \varphi\ (\varphi \in \text{Msr rlm}\ \varphi)$.

.4 mbl φ = measurable φ = $\mathsf{E}\ A \in \text{dmn}\ \varphi\ (\varphi \in$ Measure and

$$\varphi(T) = \varphi(TA) + \varphi(T{\sim}A)$$

whenever $T \in \text{dmn}\ \varphi$).

.5 mbl′ φ = mbl $\varphi \cap$ dmn′ φ.

.6 mbl″ φ = mbl $\varphi \cap$ Join″ mbl′ φ.

Of course mbl″ φ = Join″ mbl′ φ whenever $\varphi \in$ Measure.

.7 sct $\varphi\ T$ = section $\varphi\ T$ = the function ψ on dmn φ such that, for each $A \in \text{dmn}\ \varphi$,

$$\psi(A) = \varphi(TA).$$

.8 sms φ = submeasure φ = $\mathsf{E}\ \psi\ (\varphi \in$ Measure and ψ = sct $\varphi\ T$ for some $T \in \text{dmn}'\ \varphi$).

.9 Zr φ = $\mathsf{E}\ \beta\ (\varphi \in$ Measure and $\beta \in \text{zr}\ \varphi$).

.10 sb′ $\varphi\ B$ = $\mathsf{E}\ \beta \in \text{dmn}\ \varphi\ (\beta{\sim}B \in \text{Zr}\ \varphi)$.

Thus $\beta \in \text{sb}'\ \varphi\ B$ if and only if β is a subset of rlm φ and β is φ almost a subset of B.

.11 cvr $\varphi\ A\ F$ = $\mathsf{E}\ G$ (G is such a countable subfamily of F that $A \in \text{sb}'\ \varphi\ \sigma G$).

.12 $\overline{\mathsf{m}}\ f\ \varphi$ = the function ψ on dmn φ for which

$$\psi(A) = \inf G \in \text{cvr}\ \varphi\ A\ \text{dmn}\ f\ \Sigma\ \beta \in G\ f(\beta)$$

whenever $A \in \text{dmn}\ \varphi$.

Thus if $\varphi \in$ Measure, $A \subset \text{rlm}\ \varphi$, $f \in$ gauge and $\psi = \overline{\mathsf{m}}\ f\ \varphi$, then $\psi(A)$ is the infimum of numbers of the form

$$\Sigma\ \beta \in G\ f(\beta),$$

where G is a countable subfamily of dmn f covering φ almost all of A.

.13 Al $\varphi\ A\ x\ P$ if and only if $A \in \text{sb}'\ \varphi\ \mathsf{E}\ x\ P$.

.14 Alm $\varphi\ x\ P$ if and only if Al φ rlm $\varphi\ x\ P$.

In .13 and .14 we allow 'P' to be replaced by an arbitrary formula, such as for example

$$`(f(x,y) > 0)\text{'}.$$

.15 Hull H = E $\varphi \in$ Msr σH (corresponding to each $A \in$ dmn φ there is an

$$A' \in H \cap \text{sp}\ A$$

for which $\varphi(A) = \varphi(A')$).

.16 count S = the function φ on sb S for which

$$\varphi(\beta) = \Sigma\ x \in \beta\ 1$$

whenever $\beta \subset S$.

2.3 THEOREM. If $\varphi \in$ Msr S, then

$$S = \text{rlm}\ \varphi \quad \text{and} \quad \varphi \in \text{Measure}.$$

2.4 THEOREM. If $\varphi \in$ Msr S, then a necessary and sufficient condition that $A \in$ mbl φ is that

$$\psi(S) = \psi(A) + \psi(S{\sim}A)$$

whenever $\psi \in$ sms φ.

2.5 THEOREM. If

$$\varphi \in \text{Msr}\ S, \quad \psi = \text{sct}\ \varphi\ A, \quad A' = SA,$$

then:

.1 ψ = sct $\varphi\ A' \in$ Msr S;

.2 $\psi(\beta) \leqslant \varphi(\beta)$ whenever $\beta \in$ dmn φ;

.3 $\psi(S \sim A) = 0$;

.4 $A' \in \text{mbl}\ \psi$;

.5 $\text{mbl}\ \varphi \subset \text{mbl}\ \psi$;

.6 if $A \in \text{mbl}\ \varphi$, then

$$\begin{aligned}\text{mbl}\ \varphi \cap \text{sb}\ A &= \text{mbl}\ \psi \cap \text{sb}\ A,\\ \text{mbl}'\ \varphi \cap \text{sb}\ A &= \text{mbl}'\ \psi \cap \text{sb}\ A,\\ \text{mbl}''\ \varphi \cap \text{sb}\ A &= \text{mbl}''\ \psi \cap \text{sb}\ A;\end{aligned}$$

.7 if $B \subset S$, then

$$\text{cvr}\ \varphi\ (A\,B)\ H = \text{cvr}\ \psi\ B\ H.$$

2.6 THEOREM. If $\varphi \in \text{Msr}\ S$ and $\psi \in \text{sms}\ \varphi$ then:

.1 $\psi \in \text{Msr}\ S$;

.2 $\psi(S) < \infty$;

.3 $\text{mbl}\ \varphi \subset \text{mbl}\ \psi$.

2.7 THEOREM. If

.1 $A \subset A'$,

.2 $\varphi(A) = \varphi(A') < \infty$,

.3 $T \in \text{mbl}\ \varphi$,

then

$$\varphi(TA) = \varphi(TA').$$

Not well known is the

2.8 THEOREM. If

.1 $f \in \text{Hull mbl}\ \varphi$,

.2 $\text{Zr}\ \varphi \subset \text{Zr}\ f$,

.3 $A \subset A'$,

.4 $\varphi(A) = \varphi(A') < \infty$,

.5 $T \in \text{mbl}\ \varphi$,

then

$$f(TA) = f(TA').$$

Proof. With the help of .1 so choose

$$A'' \in \text{mbl } \varphi$$

that

$$TA \subset A''$$

and

.6 $$f(TA) = f(A'').$$

From 2.7 we infer

$$\begin{aligned} \varphi(TA' \sim A'') &= \varphi(T \sim A''\ A') = \varphi(T \sim A''\ A) \\ &= \varphi(TA \sim A'') = \varphi(0) = 0. \end{aligned}$$

Hence according to .2

$$f(TA' \sim A'') = 0.$$

From this and .6 we conclude

$$\begin{aligned} f(TA') &\leqslant f(TA'\ A'') + f(TA' \sim A'') \\ &= f(TA'\ A'') \leqslant f(A'') = f(TA) \leqslant f(TA'). \end{aligned}$$

2.9 LEMMA. If

$$\begin{aligned} &\varphi \in \text{Msr } S, \quad \psi = \text{sct } \varphi\ A, \quad \beta \subset S, \\ &\beta A \subset A', \quad \varphi(\beta A) = \varphi(A'), \quad \beta' = AA' \cup S \sim A, \end{aligned}$$

then

$$\beta \subset \beta' \quad \text{and} \quad \psi(\beta) = \psi(\beta').$$

It is now easy to check the

2.10 THEOREM. If

$$\varphi \in \text{Hull mbl } \varphi, \;\; \psi = \text{sct } \varphi \; A,$$

then:

.1 $\psi \in$ Hull mbl ψ;

.2 if $A \in$ mbl φ then

$$\psi \in \text{Hull mbl } \varphi \subset \text{Hull mbl } \psi.$$

2.11 LEMMA. If

$$\text{Zr } \varphi \subset \text{Zr } \psi, \;\; A \in \text{sb}' \; \varphi \; B, \;\; T \in \text{dmn } \psi,$$

then

$$\psi(TA) \leqslant \psi(TB).$$

Recall that $\sigma 0 = 0$, that $\Sigma \; A \in 0 \; f(A) = 0$ and that an empty infimum is ∞, in checking the following

2.12 THEOREM. If

$$\varphi \in \text{Msr } S, \;\; f \in \text{gauge}, \;\; \overline{f} = \overline{\text{m}} \, f \; \varphi,$$

then:

.1 $\overline{f} \in$ Msr S;

.2 $\overline{f}(S\beta) \leqslant f(\beta)$ whenever $\beta \in$ dmn f;

.3 Zr $\varphi \subset$ Zr $\overline{f}$;

.4 if

$$f \in \text{Msr } S, \quad \text{Zr } \varphi \subset \text{Zr } f,$$

then

$$f = \overline{f};$$

.5 $\overline{f} = \overline{\mathsf{m}}\, f\, \varphi$;

.6 $\overline{f} = \overline{\mathsf{m}}\, f\, \overline{f}$;

.7 if

$$g \subset f, \quad \overline{g} = \overline{\mathsf{m}}\, g\, \varphi,$$

then

$$\overline{f}(\beta) \leqslant \overline{g}(\beta) \text{ whenever } \beta \subset S.$$

2.13 THEOREM. If $\varphi \in$ Measure then

$$\varphi = \overline{\mathsf{m}}\, \varphi\, \varphi.$$

2.14 THEOREM. If

$$\begin{gathered} \varphi \in \text{Measure}, \quad f \in \text{gauge}, \quad \overline{f} = \overline{\mathsf{m}}\, f\, \varphi, \\ C \in \text{dmn } \varphi, \quad A \in \text{dmn } \varphi, \quad \psi = \text{sct } \overline{f}\, C, \\ f(T) \geqslant \overline{f}(TCA) + \overline{f}(TC \sim A) \end{gathered}$$

whenever $T \in \text{dmn } f$, then

$$A \in \text{mbl } \psi.$$

Proof. Suppose

$$r > 0, \quad T' \in \text{dmn}' \, \psi,$$

note that

$$\overline{f}(T'C) < \infty,$$

and choose

$$G \in \text{cvr}\ \varphi\ (T'C)\ \text{dmn}\ f$$

so that

$$\Sigma\ \beta \in G\ f(\beta) \leqslant \overline{f}(T'C) + r.$$

Helped by 2.12.1 and 2.11 we now infer

$$\begin{aligned}\psi(T') + r = \overline{f}(T'C) + r &\geqslant \Sigma\ \beta \in G\ f(\beta)\\ &\geqslant \Sigma\ \beta \in G(\overline{f}(\beta CA) + \overline{f}(\beta C{\sim}A))\\ &= \Sigma\ \beta \in G\ \overline{f}(\beta CA) + \Sigma\ \beta \in G\ \overline{f}(\beta C{\sim}A)\\ &\geqslant \overline{f}(CA\sigma G) + \overline{f}(C{\sim}A\sigma G)\\ &\geqslant \overline{f}(T'CA) + \overline{f}(T'C{\sim}A)\\ &= \psi(T'A) + \psi(T'{\sim}A).\end{aligned}$$

Consequently

$$\psi(T') + r \geqslant \psi(T'A) + \psi(T'{\sim}A)$$

and the desired conclusion is at hand.

2.15 THEOREM. If

$$\begin{aligned}&\varphi \in \text{Measure},\ f \in \text{gauge},\ \overline{f} = \overline{\text{m}}\, f\, \varphi,\\ &\varphi_1 = \text{sct}\ \varphi\ A,\ F = \text{sct}\ \overline{f}\ A,\ \overline{f}_1 = \overline{\text{m}}\, f\, \varphi_1,\end{aligned}$$

then

$$F = \overline{f}_1 .$$

Proof. Let

$$H = \text{dmn}\, f$$

and use 2.5.7 to see that

$$\text{cvr}\ \varphi\ (\beta A)\, H = \text{cvr}\ \varphi_1\ \beta\, H$$

whenever $\beta \in \text{dmn}\ \varphi$. The conclusion follows easily.

Theorem 2.16 below is due to Trevor J. McMinn[4]. We state and prove it in a form convenient to our purposes.

2.16 THEOREM. If $\varphi \in \text{Msr}\ S$, $\varphi(S \sim \bigcup\ m \in \omega\ K(m)) = 0$, and $K(n+1) \supset K(n) \in F \subset$ mbl sct $\varphi\ K(n)$ whenever $n \in \omega$, then $F \subset \text{mbl}\ \varphi$.

Proof. We suppose $\psi_n = \text{sct}\ \varphi\ K(n)$ whenever $n \in \omega$, and divide the remainder of the proof into two parts. Notice that 'lin $n\ \psi_n\ (A)$' is our 6.5.6R[1] notation for the more usual '$\lim\limits_{n \to \infty} \psi_n\ (A)$'.

Part 1. If $A \subset S$ then $\varphi(A) = \text{lin}\ n\ \psi_n\ (A)$.

Proof. Assume without loss of generality that $K(0) = 0$ and suppose that $n' = \mathrm{E}\ m \in \omega$ $(0 \leqslant m < n)$ whenever $n \in \omega$. Then

$$\begin{aligned}
\varphi(A) &\leqslant \Sigma\ n \in \omega\ \varphi(A \cap [K(n+1) \sim K(n)]) \\
&= \text{lin}\ n\ \Sigma\ m \in n'\ \varphi(A \cap [K(m+1) \sim K(m)]) \\
&= \text{lin}\ n\ \Sigma\ m \in n'\ \psi_n\ (A \cap [K(m+1) \sim K(m)]) \\
&= \text{lin}\ n\ \psi_n\ (A \cap \bigcup\ m \in n'\ [K(m+1) \sim K(m)]) \\
&= \text{lin}\ n\ \psi_n\ (A \cap K(n)) \\
&= \text{lin}\ n\ \psi_n\ (A) \\
&\leqslant \varphi(A).
\end{aligned}$$

Part 2. $F \subset \text{mbl}\ \varphi$.

Proof. Suppose $B \in F$ and $T \subset S$. Use Part 1 to see that

$$\begin{aligned}\varphi(T) &= \text{lin}\ n\ \psi_n\,(T)\\ &= \text{lin}\ n\ (\psi_n\,(TB) + \psi_n\,(T\sim B))\\ &= \text{lin}\ n\ \psi_n\,(TB) + \text{lin}\ n\ \psi_n\,(T\sim B)\\ &= \varphi(TB) + \varphi(T\sim B).\end{aligned}$$

Hence $B \in \text{mbl}\ \varphi$ and $F \subset \text{mbl}\ \varphi$.

3. WEBS.

3.1 DEFINITIONS.

.1 strc $R\ A = R \cap \mathsf{E}\, x,y\ (x \in A)$.

.2 strn $R\ A = R \cap \mathsf{E}\, x,y\ (y \in A)$.

3.2 DEFINITIONS.

.1 web = $\mathsf{E}\ W$ (W is such a function that $W(x)$ is a run whenever $x \in$ dmn W).

.2 knap $W\ F = \mathsf{E}\ x \in$ dmn W ($W \in$ web and $W(x)$ is frequently in F).

.3 knit $W\ F = \mathsf{E}\ x \in$ dmn W ($W \in$ web and $W(x)$ is eventually in F).

.4 cwb $W\ F$ = the function W' on knap $W\ F$ for which

$$W'(x) = \text{strn } W(x)\ F$$

whenever $x \in$ knap $W\ F$.

.5 coweb $W = \mathsf{E}\ W' \in$ web ($W \in$ web and $W' \subset$ cwb $W\ F$ for some family F).

.6 cob $W\ B = \mathsf{E}\ W' \in$ coweb W (dmn $W' \subset B$).

.7 Webalm $\varphi\ W\ F$ if and only if $W \in$ web and

$$\text{knap } W \sim F \in \text{Zr } \varphi.$$

.8 Weball $W\ F$ if and only if $W \in$ web and

$$\text{knap } W \sim F = 0.$$

.9 reach W = rng σ rng W.

We shall think of $W(x)$ in 3.2.1 as running in a family of sets. Thus if $W \in$ web then: reach W is the smallest family F such that $W(x)$ runs in F whenever $x \in$ dmn W; reach W is of prime importance to us since it is made up of the sets used in forming the quotients which approximate the derivative.

We make use of 6.5R$^{(1)}$ in

3.3 DEFINITIONS.

.1 $\overline{\mathsf{D}}\ W\ f\ g\ x = \overline{\mathrm{lm}}\ B\ W(x) \dfrac{f(B)}{g(B)}$

.2 $\underline{\mathsf{D}}\ W\ f\ g\ x = \underline{\mathrm{lm}}\ B\ W(x) \dfrac{f(B)}{g(B)}$

.3 $\mathsf{D}\ W\ f\ g\ x = \mathrm{lm}\ B\ W(x) \dfrac{f(B)}{g(B)}$

If W is a web, f and g are real valued functions defined on reach W, and $x \in$ dmn W, then $\mathsf{D}\ W\ f\ g\ x$ is the W derivative of f with respect to g at x.

3.4 THEOREMS.

.1 If R is frequently in A and

$$R' = \mathrm{strn}\ R\ A\ ,$$

then

$$R' \text{ is a corun of } R,$$
$$\mathrm{dmn}\ R' = \mathrm{dmn}\ R,$$
$$\mathrm{rng}\ R' = A \cap \mathrm{rng}\ R.$$

.2 strn (strn R A) B = strn R $(A \cap B)$.

.3 If R is a relation and rng $R \subset A$ then

$$\mathrm{strn}\ R\ A = R\ .$$

Theorems 3.5 below are dual in the sense that theorems are formed from them by interchanging 'knit' and 'knap', '$\cap$' and '$\cup$', '$\subset$' and '$\supset$'.

3.5 THEOREMS.

.1 knit W 0 = 0 = knap W 0.

.2 If $W \in$ web then

$$\mathrm{knap}\ W \sim F = \mathrm{dmn}\ W \sim \mathrm{knit}\ W\ F.$$

.3 If $W \in$ web then

$$\text{knit } W \sim F = \text{dmn } W \sim \text{knap } W\ F.$$

.4 knit $W\ (F \cap G) =$ knit $W\ F \cap$ knit $W\ G$.

.5 knap $W\ (F \cup G) =$ knap $W\ F \cup$ knap $W\ G$.

.6 If $G \subset F$ then knit $W\ G \subset$ knit $W\ F$.

.7 If $G \subset F$ then knap $W\ G \subset$ knap $W\ F$.

.8 knit $W\ F \subset$ knap $W\ F$.

3.6 THEOREM. If $W \in$ web then

$$\text{reach } W = \bigcup x \in \text{dmn } W \text{ rng } W(x).$$

3.7 THEOREM. If $W \in$ web and $V =$ cwb $W\ F$ then

.1 dmn $V =$ knap $W\ F \subset$ dmn W,

.2 reach $V \subset F \cap$ reach W,

.3 $V(x) =$ strn $W(x)\ F$ whenever $x \in$ dmn V,

.4 $V \in$ coweb W.

3.8 THEOREMS.

.1 knit $W\ F \subset$ knap $W\ F \subset$ dmn W.

.2 If $W \in$ web and reach $W \subset F$ then

$$\text{dmn } W = \text{knit } W\ F = \text{knap } W\ F \text{ and } W = \text{cwb } W\ F.$$

.3 If $W' \subset W \in$ web then $W' \in$ coweb W.

.4 If $W' \in$ coweb W then

$$W'(x) \text{ is a corun of } W(x)$$

whenever $x \in$ dmn W'.

3.9 THEOREM. If $W' \in \text{coweb } W$ then

.1 $\text{dmn } W' \subset \text{dmn } W$,

.2 $\text{reach } W' \subset \text{reach } W$,

.3 $\text{knap } W' F \subset \text{knap } W F$.

3.10 LEMMA. If

$$W \in \text{web}, \quad W' \subset V = \text{cwb } W F,$$
$$W'' \subset V' = \text{cwb } W' G, \quad V'' = \text{cwb } W (F \cap G),$$

then

$$W'' \subset V''.$$

Proof. With the help of 3.8.3, 3.9.2 and 3.7.2 we infer

$$\begin{aligned}\text{reach } W'' &\subset \text{reach } V' \subset G \cap \text{reach } W' \\ &\subset G \cap \text{reach } V \subset G \cap F \cap \text{reach } W \\ &\subset F \cap G.\end{aligned}$$

Thus because of 3.8.2, 3.8.3, 3.9.3, 3.7.4 and 3.7.1 we have

$$\begin{aligned}\text{dmn } W'' = \text{knap } W'' (F \cap G) &\subset \text{knap } V' (F \cap G) \\ &\subset \text{knap } W' (F \cap G) \subset \text{knap } V (F \cap G) \\ &\subset \text{knap } W (F \cap G) = \text{dmn } V''.\end{aligned}$$

Consequently

$$\text{dmn } W'' \subset \text{dmn } V''$$

and since

$$\text{dmn } W'' \subset \text{dmn } V' \subset \text{dmn } W'$$

we may use 3.7.3 and 3.4.2 in checking that if

$$x \in \operatorname{dmn} W''$$

then

$$\begin{aligned} W''(x) = V'(x) &= \operatorname{strn} W'(x)\, G \\ &= \operatorname{strn} (\operatorname{strn} W(x)\, F)\, G \\ &= \operatorname{strn} W(x)\, (F \cap G) \\ &= V''(x). \end{aligned}$$

Accordingly

$$W'' \subset V''$$

We can now easily check 3.11.1 below.

3.11 THEOREMS.

.1 If $W' \in \operatorname{coweb} W$ and $W'' \in \operatorname{coweb} W'$ then

$$W'' \in \operatorname{coweb} W.$$

.2 If $W' \in \operatorname{coweb} W$ then

$$\operatorname{coweb} W' \subset \operatorname{coweb} W$$

and

$$\operatorname{cob} W'\, B \subset \operatorname{cob} W\, B.$$

3.12 THEOREMS.

.1 If $G \subset F$ and Webalm $\varphi\ W\ G$ then

$$\text{Webalm } \varphi\ W\ F.$$

.2 If $G \subset F$ and Weball W G then

$$\text{Weball } W\ F.$$

.3 If Webalm φ W F and Webalm φ W G then

$$\text{Webalm } \varphi\ W\ (F \cap G).$$

.4 If Weball W F and Weball W G then

$$\text{Weball } W\ (F \cap G).$$

.5 If $\varphi \in$ Measure and Weball W F then

$$\text{Webalm } \varphi\ W\ F.$$

.6 If $W' \in$ coweb W and Webalm φ W F then

$$\text{Webalm } \varphi\ W'\ F.$$

.7 If $W' \in$ coweb W and Weball W F then

$$\text{Weball } W'\ F.$$

.8 If

$$\text{Webalm } \varphi\ W\ F,\ \ W' = \text{cwb } W\ F,$$
$$A = \text{dmn } W,\ \ A' = \text{dmn } W',$$

then

$$A \sim A' \in \text{Zr } \varphi.$$

Proof.

$$A \sim A' = \text{dmn } W \sim \text{knap } W\ F = \text{knit } W \sim F$$
$$\subset \text{knap } W \sim F \in \text{Zr } \varphi.$$

.9 If $W \in$ web and reach $W \subset F$ then

$$\text{Weball } W\ F.$$

Proof. Helped by 3.8.2 we conclude

$$0 = \text{dmn } W \sim \text{knit } W\ F = \text{knap } W \sim F.$$

.10 If

$$\varphi \in \text{Measure}, \quad W \in \text{web}, \quad \text{reach } W \subset F,$$

then

$$\text{Webalm } \varphi\ W\ F.$$

Proof. Use .9 and .5.

3.13 THEOREM. If

$$W_1 \in \text{web}, \quad W_2 \in \text{web}, \quad \text{dmn } W_1 \cap \text{dmn } W_2 = 0,$$

then:

.1 $W_1 \cup W_2 \in$ web;

.2 knap $(W_1 \cup W_2)\ F =$ knap $W_1\ F \cup$ knap $W_2\ F$;

.3 knit $(W_1 \cup W_2)\ F =$ knit $W_1\ F \cup$ knit $W_2\ F$.

Hereinafter it is understood that, for each a, neither $a/0$ nor ∞/∞ is either a finite or infinite real number.

3.14 THEOREM. If

$$W \in \text{web},\ \ g \in \text{gauge},\ \ h \in \text{gauge},\ \ 0 < \lambda < \infty,$$
$$M = \mathsf{E}\,\beta\,[g(\beta) < \lambda \cdot h(\beta)],\ \ N = \mathsf{E}\,\beta\,[\lambda \cdot h(\beta) < g(\beta)],$$

then:

.1 $\mathsf{E}\,x\,(\underline{\mathsf{D}}\;W\,g\,h\,x < \lambda) \subset \text{knap}\;W\,M;$

.2 $\text{knap}\;W \sim M \subset \text{dmn}\;W \sim \mathsf{E}\,x\,(\overline{\mathsf{D}}\;W\,g\,h\,x < \lambda);$

.3 $\mathsf{E}\,x\,(\overline{\mathsf{D}}\;W\,g\,h\,x > \lambda) \subset \text{knap}\;W\,N;$

.4 $\text{knap}\;W \sim N \subset \text{dmn}\;W \sim \mathsf{E}\,x\,(\underline{\mathsf{D}}\;W\,g\,h\,x > \lambda).$

3.15 THEOREM. If

$$W \in \text{web},\ \ g \in \text{gauge},\ \ h \in \text{gauge},$$
$$0 \leqslant \underline{\mathsf{D}}\;W\,g\,h\,x = \overline{\mathsf{D}}\;W\,g\,h\,x \leqslant \infty,$$

then

$$0 \leqslant \underline{\mathsf{D}}\;W\,g\,h\,x = \mathsf{D}\;W\,g\,h\,x = \overline{\mathsf{D}}\;W\,g\,h\,x \leqslant \infty.$$

3.16 THEOREM. If

$$W \in \text{web},\ \ g \in \text{gauge},\ \ h \in \text{gauge},$$

then

$$\text{dmn}\;W \sim \mathsf{E}\,x\,(0 \leqslant \underline{\mathsf{D}}\;W\,g\,h\,x \lneq \overline{\mathsf{D}}\;W\,g\,h\,x \leqslant \infty)$$
$$\subset (\text{knap}\;W \sim \text{dmn}'\,g) \cup (\text{knap}\;W \sim \text{dmn}'\,h) \cup (\text{knap}\;W\;\text{zr}\;h).$$

According to 6.1.4R[(1)] and 2.2R, $\mathsf{D}\;W\,g\,h\,x$ is always either the universe or a real number. Use is made of this fact in

3.17 THEOREM. If $W \in \text{web}$,

$$g_1(\beta) = g(\beta) \text{ and } h_1(\beta) = h(\beta)$$

whenever $\beta \in F$, then

$$\text{dmn } W \sim \mathsf{E}\, x\, (\mathsf{D}\, W\, g_1\, h_1\, x = \mathsf{D}\, W\, g\, h\, x) \subset \text{knap } W \sim F.$$

3.18 THEOREM. If

$$W \in \text{web},\ h \in \text{gauge},$$
$$0 \leqslant G(\beta) \leqslant g(\beta) \text{ whenever } \beta \in F,$$

then

$$\mathsf{E}\, x\, (\mathsf{D}\, W\, g\, h\, x = 0) \sim \mathsf{E}\, x\, (\mathsf{D}\, W\, G\, h\, x = 0) \subset \text{knap } W \sim F.$$

3.19 THEOREM. If

$$W \in \text{web},\ f \in \text{gauge},\ g \in \text{gauge},\ h \in \text{gauge},$$
$$0 \leqslant \mathsf{D}\, W\, f\, g\, x < \infty,\ 0 \leqslant \mathsf{D}\, W\, g\, h\, x < \infty,$$

then

$$\mathsf{D}\, W\, f\, h\, x = \mathsf{D}\, W\, f\, g\, x \cdot \mathsf{D}\, W\, g\, h\, x.$$

3.20 THEOREM. If

$$W \in \text{web},\ g \in \text{gauge},\ h \in \text{gauge},$$
$$0 < \mathsf{D}\, W\, g\, h\, x < \infty,$$

then

$$\mathsf{D}\, W\, h\, g\, x = 1 / \mathsf{D}\, W\, g\, h\, x.$$

4. SOME TILE THEORY.

4.1 DEFINITIONS.

.1 cwr φ W = E G ($W \in$ web) $\cap$ cvr φ dmn W reach W.

Thus if $W \in$ web then

$$\text{cwr } \varphi\ W = \text{cvr } \varphi \text{ dmn } W \text{ reach } W.$$

.2 covertileweb φ = E W (cwr φ $W \neq 0$).

.3 covertiled φ W = E B ($\varphi \in$ Measure, $W \in$ web, cob W $B \subset$ covertileweb φ).

.4 tiled f φ W = E B [$f \in$ gauge, $\varphi \in$ Measure, $W \in$ web, and corresponding to each $r > 0$ and each $W' \in$ cob W B there is a $G \in$ cwr φ W' for which

$$\Sigma\ \beta \in G\ f(\beta) \leqslant f(B) + r\,].$$

4.2 DEFINITION. covertile W = E φ (

.1 Webalm φ W dmn φ,

.2 Webalm φ W covertiled φ W,

.3 cvr φ dmn W covertiled φ $W \neq 0$).

4.3 DEFINITIONS.

.1 pretile φ W = E $f \in$ gauge ($\varphi \in$ covertile W and Webalm φ W dmn$'$ f).

.2 tile φ W = E $f \in$ pretile φ W (dmn $f \subset$ tiled f φ W).

.3 localtile φ W = E $f \in$ pretile φ W (

$$\text{Webalm } \varphi\ W \text{ tiled } f\ \varphi\ W).$$

.4 fairtile W = E φ ($\varphi \in$ tile φ W).

.5 basetile W = E φ ($\varphi \in$ localtile φ W).

In 4.3.2 we have improved on the definition of addivelous functions in Perfect Blankets[5]. If we agree that

pseudotile φ W = E $f \in$ pretile φ W [corresponding to each $B \in$ dmn f, each $W' \in$ cob W B, and each $r > 0$, there is such a countable subfamily G of reach W' that

$$\varphi(\text{dmn } W' \sim \sigma\ G) \leqslant r$$

and

$$\Sigma\ \beta \in G\ f(\beta) \leqslant f(B) + \mathrm{r}\ ]$$

then clearly

$$\text{tile}\ \varphi\ W \subset \text{pseudotile}\ \varphi\ W;$$

moreover much can be learned about pseudotile φ W by employing the construction used by C. A. Hayes in section 3 of *φ-Pseudo-strong blankets*[6]. In the interest of simplicity we have not used this approach.

An alternative approach is also suggested by the work of R. de Possel in *Dérivation abstraite des fonctions d'ensemble*[7]. Here we would associate with each x a non-vacuous *family* of runs. One reason we avoid this sort of beginning is that corresponding to it, in a natural way, is a web which generates everything of interest to us in the alternative situation.

With continuity (absolute) in mind we make

4.4 DEFINITIONS.

.1 contiled $f\ \varphi\ W = \mathsf{E}\ B \in \text{tiled}\ f\ \varphi\ W\ [$

$$f(B) \leqslant \Sigma\ \beta \in G\ f(\beta) \text{ whenever } G \in \text{cvr}\ \varphi\ B\ \text{dmn}\ f\,].$$

.2 contile $\varphi\ W = \mathsf{E}\ f \in \text{pretile}\ \varphi\ W\ ($

$$\text{dmn}\ f \subset \text{contiled}\ f\ \varphi\ W).$$

4.5 THEOREMS.

.1 tiled $f\ \varphi\ W \subset$ covertiled $\varphi\ W$.

.2 contiled $f\ \varphi\ W \subset$ tiled $f\ \varphi\ W \subset$ dmn f.

.3 contile $\varphi\ W \subset$ tile $\varphi\ W \subset$ localtile $\varphi\ W \subset$ pretile $\varphi\ W \subset$ gauge.

.4 fairtile $W \subset$ basetile $W \subset$ covertile $W \subset$ Measure.

.5 If $\varphi \in$ covertile W then

$$\varphi \in \text{Measure}, \quad W \in \text{web}, \quad \text{dmn } W \in \text{dmn } \varphi.$$

Proof. In the light of .4, 4.2.1 and 3.2.7 it is clear that $\varphi \in$ Measure and $W \in$ web. If cvr φ A $F \neq 0$ then, according to 2.2.11 and 2.2.10,

$$A \in \text{dmn } \varphi.$$

Because of this and 4.2.3

$$\text{dmn } W \in \text{dmn } \varphi.$$

With the help of 3.11 and 3.12 it is fairly easy to check the

4.6 THEOREM. If $W' \in$ coweb W then:

.1 covertiled φ $W \subset$ covertiled φ W';

.2 tiled f φ $W \subset$ tiled f φ W';

.3 covertile $W \subset$ covertile W';

.4 pretile φ $W \subset$ pretile φ W';

.5 tile φ $W \subset$ tile φ W'.

Akin to 4.6 and easy to check is

4.7 THEOREM. If

$$\text{Zr } \varphi \subset \text{Zr } \psi, \quad \text{dmn } \varphi \subset \text{dmn } \psi,$$

then:

.1 sb′ φ $A \subset$ sb′ ψ A;

.2 cvr φ A $F \subset$ cvr ψ A F;

.3 cwr φ $W \subset$ cwr ψ W;

.4 covertileweb $\varphi \subset$ covertileweb ψ;

.5 covertiled φ $W \subset$ covertiled ψ W;

.6 tiled f φ $W \subset$ tiled f ψ W;

.7 if $\varphi \in$ covertile W then $\psi \in$ covertile W;

.8 pretile φ $W \subset$ pretile ψ W;

.9 tile φ $W \subset$ tile ψ W.

We use Definition 3.1.1 in formulating

4.8 LEMMA. If

$$W \in \text{web}, \quad A = \text{dmn } W, \quad H \in \text{cvr } \varphi \; A \; F,$$

K is such a function on H that

$$K(\beta) \in \text{cwr } \varphi \text{ strc } W \; \beta$$

whenever $\beta \in H$, and finally

$$G = \bigcup \beta \in H \; K(\beta),$$

then

$$G \in \text{cwr } \varphi \; W.$$

Proof. Notice that

$$\beta \; A = \text{dmn strc } W \; \beta \in \text{sb}' \; \varphi \; \sigma(K(\beta))$$

whenever $\beta \in H$. Accordingly

$$\begin{aligned} A \;\; \sigma H \sim \sigma G &= \bigcup \beta \in H \; (\beta \; A) \sim \bigcup \beta \in H \; \sigma(K(\beta)) \\ &\subset \bigcup \beta \in H \; [\beta \; A \sim \sigma(K(\beta))] \in \text{Zr } \varphi, \end{aligned}$$

and we are now sure that

$$A \;\; \sigma H \sim \sigma G \in \text{Zr } \varphi$$

and hence that

$$A \sim \sigma G \subset A \sim \sigma H \cup A \ \sigma H \sim \sigma G \in \mathrm{Zr}\ \varphi.$$

Thus

$$A \in \mathrm{sb}'\ \varphi\ \sigma G$$

and, since

$$K(\beta) \subset \text{reach strc } W\ \beta \subset \text{reach } W$$

whenever $\beta \in H$, it is now clear that

$$G \subset \text{reach } W.$$

Thus

$$G \in \mathrm{cvr}\ \varphi\ A \text{ reach } W = \mathrm{cwr}\ \varphi\ W$$

and the proof is complete.

4.9 LEMMA. If

$$W \in \text{web},\ \ A = \text{dmn } W,\ \ H \in \mathrm{cvr}\ \varphi\ A\ F,$$
$$f \in \text{gauge},\ \ p \in \text{gauge},$$

K is such a function on H that

$$K(\beta) \in \mathrm{cwr}\ \varphi \text{ strc } W\ \beta$$

and

$$\Sigma\ \alpha \in K(\beta)\ f(\alpha) \leqslant f(\beta) + p(\beta)$$

whenever $\beta \in H$, and finally

$$G = \bigcup \beta \in H\, K(\beta),$$

then

$$G \in \text{cwr}\ \varphi\ W$$

and

$$\Sigma\ \beta \in G\ f(\beta) \leqslant \Sigma\ \beta \in H\ f(\beta) + \Sigma\ \beta \in H\ p(\beta).$$

Proof. Lemma 4.8 assures us

$$G \in \text{cwr}\ \varphi\ W.$$

Moreover

$$\begin{aligned} \Sigma\ \beta \in G\ f(\beta) &= \Sigma\ \alpha \in G\ f(\alpha) \\ &\leqslant \Sigma\ \beta \in H\ \Sigma\ \alpha \in K(\beta)\ f(\alpha) \\ &\leqslant \Sigma\ \beta \in H\ (f(\beta) + p(\beta)) \\ &= \Sigma\ \beta \in H\ f(\beta) + \Sigma\ \beta \in H\ p(\beta). \end{aligned}$$

Very useful is

4.10 LEMMA. If Webalm $\varphi\ W\ F$ and $W' = \text{cwb}\ W\ F$ then

$$\text{cwr}\ \varphi\ W' \subset \text{sb}\ F \cap \text{cwr}\ \varphi\ W.$$

Proof. Let

$$A = \text{dmn}\ W,\ \ A' = \text{dmn}\ W'$$

and use 3.12.8 to learn that

.1 $$A \sim A' \in \mathrm{Zr}\ \varphi.$$

Now if

$$G \in \mathrm{cwr}\ \varphi\ W'$$

then: because of 3.7.2

$$G \subset \mathrm{reach}\ W' \subset F \quad \text{and} \quad G \in \mathrm{sb}\ F;$$

because of .1

$$A \sim \sigma G \ \subset A \sim A' \cup A' \sim \sigma G \ \in \mathrm{Zr}\ \varphi$$

and hence

$$A \in \mathrm{sb}'\ \varphi\ \ \sigma G;$$

accordingly

$$G \in \mathrm{sb}\ F \cap \mathrm{cwr}\ \varphi\ W.$$

The desired conclusion is at hand.

4.11 THEOREM. If $C \in \mathrm{sb}'\ \varphi\ \mathrm{dmn}\ W$ and $W' = \mathrm{strc}\ W\ C$ then

$$\mathrm{sb}\ F \cap \mathrm{cwr}\ \varphi\ W' \subset \mathrm{cvr}\ \varphi\ C\ F.$$

Proof. Let

$$A = \mathrm{dmn}\ W, \ \ A' = \mathrm{dmn}\ W'$$

and note that

$$A' = A \cap C.$$

Now if

$$G \in \text{sb } F \cap \text{cwr } \varphi \; W'$$

then:

$$\begin{aligned} &G \text{ is a countable subfamily of } F; \\ &C \sim \sigma G \subset C \sim A' \cup A' \sim \sigma G \\ &\qquad\qquad = C \sim A \cup A' \sim \sigma G \in \text{Zr } \varphi; \\ &C \in \text{sb}' \; \varphi \;\; \sigma G; \end{aligned}$$

accordingly

$$G \in \text{cvr } \varphi \; C \; F.$$

The desired conclusion is at hand.

4.12 THEOREM. If $\varphi \in$ covertile W then

$$W \in \text{covertileweb } \varphi.$$

Proof. Let

$$A = \text{dmn } W, \;\; F = \text{covertiled } \varphi \; W$$

and use 4.2.3 to secure

$$H \in \text{cvr } \varphi \; A \; F.$$

Since

$$\text{strc } W \, \beta \in \text{cob } W \, \beta \subset \text{covertileweb } \varphi$$

whenever $\beta \in H$, we use 4.1.2 to secure such a function K on H that

$$K(\beta) \in \text{cwr}\ \varphi\ \text{strc}\ W\ \beta$$

whenever $\beta \in H$. Let

$$G = \bigcup\ \beta \in H\ K(\beta)$$

and use 4.8 and 4.1.2 to conclude

$$G \in \text{cwr}\ \varphi\ W, \quad W \in \text{covertileweb}\ \varphi.$$

4.13 THEOREM. If

$$\varphi \in \text{covertile}\ W \quad \text{and} \quad \text{Webalm}\ \varphi\ W\ F$$

then

$$\text{sb}\ F \cap \text{cwr}\ \varphi\ W \neq 0.$$

Proof. Let

$$W' = \text{cwb}\ W\ F.$$

Because of 4.6.3 we know

$$\varphi \in \text{covertile}\ W'.$$

We now conclude from 4.12, 4.1.2 and 4.10 that

$$0 \neq \text{cwr}\ \varphi\ W' \subset \text{sb}\ F \cap \text{cwr}\ \varphi\ W.$$

4.14 THEOREM. If

$$f \in \text{pretile } \varphi\ W, \quad F \subset \text{tiled } f\ \varphi\ W,$$
$$\text{Webalm } \varphi\ W\ F, \quad g = \text{strc } f\ F,$$

then

$$\text{dmn } g = \text{F} \quad \text{and} \quad g \in \text{tile } \varphi\ W.$$

Proof. Clearly dmn $g = F$ and because of 3.12.3 it is easy to check that $g \in$ pretile φ W. The desired conclusion is now a consequence of 4.3.2, 4.1.4 and the

Statement. If

$$B \in F, \quad V \in \text{cob } W\ B, \quad r > 0,$$

then there is a $G \in$ cwr φ V for which

$$\Sigma\ \beta \in G\ g(\beta) \leqslant g(B) + r.$$

Proof. Let

$$V' = \text{cwb } V\ F.$$

Because of 3.12.6 we are sure that

$$\text{Webalm } \varphi\ V'\ F.$$

Also, since

$$B \in \text{tiled } f\ \varphi\ W \quad \text{and} \quad V' \in \text{cob } W\ B$$

we can use 4.1.4 and 4.10 to find such a

$$G \in \text{sb } F \cap \text{cwr } \varphi \, V$$

that

$$\Sigma \, \beta \in G \, g(\beta) = \Sigma \, \beta \in G \, f(\beta) \leqslant f(B) + r = g(B) + r.$$

4.15 THEOREM. If

$$f \in \text{tile } \varphi \, W, \ \overline{f} = \overline{\text{m}} \, f \, \varphi, \ A = \text{dmn } W, \ r > 0,$$

then there is a $G \in \text{cwr } \varphi \, W$ for which

$$\Sigma \, \beta \in G \, f(\beta) \leqslant \overline{f}(A) + r.$$

Proof. Let

$$F = \text{dmn}' \, f$$

and note that

$$\text{Webalm } \varphi \, W \, F \ \text{ and } \ A \subset \text{rlm } \varphi = \text{rlm } \overline{f}.$$

If $\overline{f}(A) = \infty$ we use 4.13 to discover such a

$$G \in \text{sb } F \cap \text{cwr } \varphi \, W$$

that

$$\Sigma \, \beta \in G \, f(\beta) \leqslant \infty = \overline{f}(A) + r.$$

We now assume

$$\overline{f}(A) < \infty.$$

We use 2.2.12 to secure

$$H \in \operatorname{cvr} \varphi\, A \ \operatorname{dmn} f$$

so that

.1 $$\Sigma\, \beta \in H\, f(\beta) \leq \overline{f}\,(A) + r/2.$$

Next let p be such a positive valued function that

.2 $$\Sigma\, \beta \in H\; p\,(\beta) \leq r/2.$$

Since 4.3.2 assures us

$$\beta \in \operatorname{tiled} f\ \varphi\ W \text{ whenever } \beta \in H$$

and since

$$\operatorname{strc}\ W\,\beta \in \operatorname{cob}\ W\,\beta \text{ whenever } \beta \in H,$$

we use 4.1.4 to secure such a function K on H that for each $\beta \in H$,

$$K(\beta) \in \operatorname{cwr} \varphi \operatorname{strc}\ W\,\beta$$

and

$$\Sigma\, \alpha \in K(\beta)\, f(\alpha) \leq f(\beta) + p\,(\beta).$$

Now if

$$G = \cup\, \beta \in H\ K\,(\beta)$$

then from .1, .2 and 4.9 we can conclude

$$G \in \operatorname{cwr} \varphi\ W$$

and

$$\Sigma\ \beta \in G\ f(\beta) \leqslant \Sigma\ \beta \in H\ f(\beta) + \Sigma\ \beta \in H\ p(\beta)$$
$$\leqslant \overline{f}(A) + r/2 + r/2 = \overline{f}(A) + r.$$

4.16 THEOREM. If

$$\text{Webalm}\ \varphi\ W\ F,\ \ f \in \text{tile}\ \varphi\ W,$$
$$\overline{f} = \overline{\text{m}}\,f\ \varphi,\ \ A = \text{dmn}\ W,\ \ r > 0,$$

then there is a $G \in \text{sb}\ F \cap \text{cwr}\ \varphi\ W$ for which

$$\Sigma\ \beta \in G\ f(\beta) \leqslant \overline{f}(A) + r.$$

Proof. Let

$$W' = \text{cwb}\ W\ F,\ \ A' = \text{dmn}\ W',$$

and note that

$$A' \subset A.$$

Because of 4.6.5 we know

$$f \in \text{tile}\ \varphi\ W'.$$

We use 4.15 and 4.10 to secure

$$G \in \text{sb}\ F \cap \text{cwr}\ \varphi\ W$$

so that

$$\Sigma\ \beta \in G\ f(\beta) \leqslant \overline{f}(A') + r \leqslant \overline{f}(A) + r.$$

4.17 THEOREM. If

$$\text{Webalm } \varphi\ W\ F,\ \ f \in \text{tile } \varphi\ W,$$
$$\overline{f} = \overline{\mathrm{m}}\ f\ \varphi,\ \ C \in \text{sb}'\ \varphi \text{ dmn } W,\ \ r > 0,$$

then there is a $G \in \text{cvr } \varphi\ C\ F$ for which

$$\Sigma\ \beta \in G\ f(\beta) \leqslant \overline{f}\,(C) + r.$$

Proof. Let

$$W' = \text{strc } W\ C,\ \ A' = \text{dmn } W',$$

and notice that

$$A' \subset C \in \text{dmn } \varphi.$$

Since, according to 4.6.5 and 3.12.6,

$$f \in \text{tile } \varphi\ W' \quad \text{and} \quad \text{Webalm } \varphi\ W'\ F,$$

we use 4.16 and 4.11 to secure

$$G \in \text{cvr } \varphi\ C\ F$$

so that

$$\Sigma\ \beta \in G\ f(\beta) \leqslant \overline{f}\,(A') + r \leqslant \overline{f}\,(C) + r.$$

Because of 4.15, 4.6.5 and 2.12.2, it is easy to check the

4.18 THEOREM. If

$$f \in \text{pretile } \varphi\ W,\ \ \overline{f} = \overline{\mathrm{m}}\ f\ \varphi,$$

then a necessary and sufficient condition that

$$f \in \text{tile } \varphi \ W$$

is that corresponding to each $W' \in \text{coweb } W$ and each $r > 0$ there is a $G \in \text{cwr } \varphi \ W'$ for which

$$\Sigma \ \beta \in G \ f(\beta) \leqslant \overline{f}\,(\text{dmn } W') + r.$$

4.19 LEMMA. If

$$f \in \text{tile } \varphi \ W, \ \ \overline{f} = \overline{\mathsf{m}}\, f \ \varphi,$$
$$A = \text{dmn } W, \ \ r > 0,$$

then there is a $G \in \text{cwr } \varphi \ W$ for which

$$\Sigma \ \beta \in G \ \overline{f}\,(\beta) \leqslant \overline{f}\,(A) + r.$$

Proof. Let

$$F = \text{dmn } \varphi$$

and use 4.2.1 to see that

$$\text{Webalm } \varphi \ W \ F.$$

Now use 4.16 to secure such a

$$G \in \text{sb } F \cap \text{cwr } \varphi \ W$$

that, in accordance with 2.12.2,

$$\overline{f}\,(A) + r \geqslant \Sigma \ \beta \in G \ f(\beta) \geqslant \Sigma \ \beta \in G \ \overline{f}\,(\beta).$$

4.20 THEOREM. If

$$f \in \text{tile } \varphi\ W,\ \overline{f} = \overline{\mathrm{m}}\, f\, \varphi,$$

then

$$\overline{f} \in \text{contile } \varphi\ W.$$

Proof. Let

$$g = \overline{f} \quad \text{and} \quad \overline{g} = \overline{\mathrm{m}}\, g\, \varphi.$$

From 2.12.5 we learn

.1 $$\overline{g} = g.$$

With the help of 4.3.1, 4.2.1 and 3.12.3 we learn

$$\text{Webalm } \varphi\ W\ (\text{dmn}'\, f \cap \text{dmn } \varphi).$$

Hence, because of 2.12.2,

$$\text{Webalm } \varphi\ W\ \text{dmn}'\, g$$

and

$$g \in \text{pretile } \varphi\ W.$$

Now for each $r > 0$ and each $W' \in \text{coweb } W$ we infer from 4.19, 4.6.5 and .1 that there is such a $G \in \text{cwr } \varphi\ W'$ that

$$\Sigma\ \beta \in G\ g(\beta) \leqslant g(\text{dmn } W') + r = \overline{g}\,(\text{dmn } W') + r.$$

That

$$g \in \text{tile } \varphi \ W$$

now follows from 4.18. From .1, 2.12 and 4.4 we conclude

$$\overline{f} = g \in \text{contile } \varphi \ W.$$

4.21 THEOREM. If

$$f \in \text{tile } \varphi \ W, \ \overline{f} = \overline{\mathrm{m}} f \varphi, \ A \in \text{dmn } \varphi,$$

then a necessary and sufficient condition that

$$A \in \text{contiled } f \ \varphi \ W$$

is that

$$f(A) = \overline{f}(A).$$

4.22 THEOREM. If

$$f \in \text{tile } \varphi \ W, \ \overline{f} = \overline{\mathrm{m}} f \varphi,$$

then a necessary and sufficient condition that

$$f \in \text{contile } \varphi \ W$$

is that

$$f(\beta) = \overline{f}(\beta) \text{ whenever } \beta \in \text{dmn } f \cap \text{dmn } \varphi.$$

We now have at once

4.23 THEOREM. If

$$f \in \text{tile } \varphi\ W, \quad \text{dmn } f = \text{dmn } \varphi, \quad \overline{f} = \overline{\mathsf{m}}\, f\, \varphi,$$

then a necessary and sufficient condition that

$$f \in \text{contile } \varphi\ W$$

is that

$$f = \overline{f}.$$

4.24 THEOREM. If

$$f \in \text{Msr rlm } \varphi \cap \text{tile } \varphi\ W,$$

then a necessary and sufficient condition that

$$f \in \text{contile } \varphi\ W$$

is that

$$\text{Zr } \varphi \subset \text{Zr } f.$$

4.25 LEMMA. If

$$\begin{aligned}
&0 < a < \infty, \quad 0 < b < \infty,\\
&\mu \in \text{gauge}, \quad \nu \in \text{tile } \varphi\ W,\\
&\overline{\mu} = \overline{\mathsf{m}}\, \mu\, \varphi, \quad \overline{\nu} = \overline{\mathsf{m}}\, \nu\, \varphi,\\
&F = \mathsf{E}\, \beta\, [a \cdot \mu(\beta) < b \cdot \nu(\beta)]\,,\\
&B \in \text{sb}'\ \varphi \text{ knap } W\ F,
\end{aligned}$$

then

$$a \cdot \overline{\mu}(B) \leqslant b \cdot \overline{\nu}(B).$$

Proof. Let

$$W' = \text{cwb } W\, F$$

and note that

$$\nu \in \text{tile } \varphi\, W'.$$

According to 3.7,

$$B \in \text{sb}'\, \varphi \text{ dmn } W' \quad \text{and} \quad \text{reach } W' \subset F.$$

Hence, if $r > 0$, we can use 3.12.10 and 4.17 to secure

$$G \in \text{cvr } \varphi\, B\, F$$

so that:

$$\begin{aligned} \overline{\nu}(B) + r/b &\geqslant \Sigma\, \beta \in G\ \nu(\beta); \\ b \cdot \overline{\nu}(B) + r &\geqslant \Sigma\, \beta \in G\ (b \cdot \nu(\beta)) \\ &\geqslant \Sigma\, \beta \in G\ (a \cdot \mu(\beta)) = a \cdot \Sigma\, \beta \in G\ \mu(\beta) \\ &\geqslant a \cdot \overline{\mu}(B). \end{aligned}$$

Hence

$$a \cdot \overline{\mu}(B) \leqslant b \cdot \overline{\nu}(B).$$

4.26 THEOREM. If

$$0 < \lambda < \infty,$$
$$g \in \text{gauge}, \quad h \in \text{tile } \varphi\, W,$$

$$\overline{g} = \overline{\mathsf{m}}\, g\, \varphi, \quad \overline{h} = \overline{\mathsf{m}}\, h\, \varphi,$$
$$\text{Al}\ \varphi\ B\ x\ (\underline{\mathsf{D}}\ W\ g\ h\ x < \lambda),$$

then

$$\overline{g}(B) \leqslant \lambda \cdot \overline{h}(B).$$

Proof. Let

$$a = 1, \quad b = \lambda,$$
$$\mu = g, \quad \nu = h,$$
$$\overline{\mu} = \overline{\mathsf{m}}\, \mu\, \varphi, \quad \overline{\nu} = \overline{\mathsf{m}}\, \nu\, \varphi,$$
$$F = \mathsf{E}\ \beta\ [a \cdot \mu(\beta) < b \cdot \nu(\beta)]$$
$$M = \mathsf{E}\ \beta\ [g(\beta) < \lambda \cdot h(\beta)].$$

Clearly $M = F$ and hence according to 3.14.1 and our hypotheses,

$$B \in \text{sb}'\ \varphi\ \text{knap}\ W\ F.$$

Since

$$\overline{\mu} = \overline{g}, \quad \overline{\nu} = \overline{h},$$

the desired conclusion now follows from 4.25.

4.27 THEOREM. If

$$0 < \lambda < \infty,$$
$$g \in \text{tile}\ \varphi\ W, \quad h \in \text{gauge},$$
$$\overline{g} = \overline{\mathsf{m}}\, g\, \varphi, \quad \overline{h} = \overline{\mathsf{m}}\, h\, \varphi,$$
$$\text{Al}\ \varphi\ B\ x\ (\overline{\mathsf{D}}\ W\ g\ h\ x > \lambda),$$

then

$$\overline{g}(B) \geqslant \lambda \cdot \overline{h}(B).$$

Proof. Let

$$\begin{aligned}
&a = \lambda, \quad b = 1, \\
&\mu = h, \quad \nu = g, \\
&\overline{\mu} = \overline{\mathsf{m}}\, \mu\, \varphi, \quad \overline{\nu} = \overline{\mathsf{m}}\, \nu\, \varphi, \\
&F = \mathsf{E}\, \beta\, [a \cdot \mu(\beta) < b \cdot \nu(\beta)], \\
&N = \mathsf{E}\, \beta\, [\lambda \cdot h(\beta) < g(\beta)].
\end{aligned}$$

Clearly $N = F$ and hence according to 3.14.3 and our hypotheses,

$$B \in \mathrm{sb}'\, \varphi\, \mathrm{knap}\, W\, F.$$

Since

$$\overline{\mu} = \overline{h}, \quad \overline{\nu} = \overline{g},$$

the desired conclusion now follows from 4.25.

From 4.26 and 4.27 we readily see

4.28 THEOREMS.

.1 If

$$\begin{aligned}
&0 < \lambda < \infty, \\
&g \in \mathrm{gauge}, \quad h \in \mathrm{tile}\, \varphi\, W, \\
&\overline{g} = \overline{\mathsf{m}}\, g\, \varphi, \quad \overline{h} = \overline{\mathsf{m}}\, h\, \varphi, \\
&\mathrm{Al}\, \varphi\, B\, x\, (\underline{\mathsf{D}}\, W\, g\, h\, x \leqslant \lambda),
\end{aligned}$$

then

$$\overline{g}(B) \leqslant \lambda \cdot \overline{h}(B).$$

.2 If

$$\begin{aligned}&0 < \lambda < \infty,\\ &g \in \text{tile } \varphi\ W,\ \ h \in \text{gauge},\\ &\overline{g} = \overline{\mathsf{m}}\ g\ \varphi,\ \ \overline{h} = \overline{\mathsf{m}}\ h\ \varphi,\\ &\text{Al } \varphi\ B\ x\ (\overline{\mathsf{D}}\ W\ g\ h\ x \geqslant \lambda),\end{aligned}$$

then

$$\overline{g}(B) \geqslant \lambda \cdot \overline{h}(B).$$

Because of 4.13 it is easy to verify the useful

4.29 LEMMA. If

$$\begin{aligned}&\varphi \in \text{covertile } W,\ \ \text{Zr } \varphi \subset \text{Zr } \psi,\\ &\text{Webalm } \varphi\ W\ F,\ \ C \in \text{sb}'\ \varphi \text{ dmn } W,\end{aligned}$$

then:

.1 if

$$\psi(\beta C) = 0 \text{ whenever } \beta \in F$$

then

$$\psi(C) = 0;$$

.2 if

$$\beta C \in \text{mbl } \psi \text{ whenever } \beta \in F$$

then

$$C \in \text{mbl}\,'\psi.$$

4.30 THEOREM. If

$$h \in \text{tile } \varphi\, W,\ \ g \in \text{gauge},\ \ \overline{g} = \overline{\mathrm{m}}\, g\, \varphi,$$

and

$$\text{Al } \varphi\, C\, x\ (\underline{\mathrm{D}}\ W\, g\, h\, x = 0),$$

then

$$\overline{g}(C) = 0.$$

Proof. Let

$$\psi = \overline{g} \ \text{ and } \ F = \text{dmn}'\, h.$$

Suppose

$$\beta \in F \ \text{ and } \ 0 < \lambda < \infty.$$

According to 4.28.1 and 2.12.2,

$$\psi(\beta C) = \overline{g}(\beta C) \leqslant \lambda \cdot \overline{h}(\beta C) \leqslant \lambda \cdot h(\beta) < \infty.$$

The arbitrary nature of λ assures us that

$$\psi(\beta C) = 0.$$

Application of 4.29.1 completes the proof.

5. DERIVATIVES AND DENSITIES.

5.1 LEMMA. If

$$g \in \text{pretile } \varphi\, W,\;\; h \in \text{pretile } \varphi\, W,\;\; \overline{h} = \overline{\mathsf{m}}\, h\, \varphi,$$

then

$$\text{Al } \overline{h} \text{ dmn } W\, x\; (0 \leqslant \underline{\mathsf{D}}\, W\, g\, h\, x \leqslant \overline{\mathsf{D}}\, W\, g\, h\, x \leqslant \infty).$$

Proof. According to 3.16,

.1 $$\text{dmn } W \sim \mathsf{E}\, x\; (0 \leqslant \underline{\mathsf{D}}\, W\, g\, h\, x \leqslant \overline{\mathsf{D}}\, W\, g\, h\, x \leqslant \infty)$$
$$\subset (\text{knap } W \sim \text{dmn}'\, g) \cup (\text{knap } W \sim \text{dmn}'\, h) \cup (\text{knap } W \text{ zr } h).$$

According to 4.3.1,

.2 $$\text{knap } W \sim \text{dmn}'\, g \in \text{Zr } \varphi$$

and

.3 $$\text{knap } W \sim \text{dmn}'\, h \in \text{Zr } \varphi.$$

Let

$$W' = \text{cwb } W \text{ zr } h$$

and use 4.6.3, 3.7.2 and 3.12.10 to check that

$$\varphi \in \text{covertile } W' \;\text{ and }\; \text{Webalm } \varphi\, W' \text{ zr } h.$$

Use 4.13 to extract

$$G \in \text{sb zr } h \cap \text{.cwr } \varphi\, W',$$

notice that

$$\text{knap } W \text{ zr } h = \text{dmn } W' = \text{dmn } W' \sim \sigma G \cup \text{dmn } W' \ \ \sigma G\,,$$

and infer with the aid of 2.12 and 4.5.5 that

$$\begin{aligned}\overline{h}\,(\text{knap } W \text{ zr } h) &\leqslant \overline{h}\,(\text{dmn } W' \sim \sigma G) + \overline{h}\,(\text{dmn } W' \ \ \sigma G) \\ &\leqslant 0 + \Sigma\, \beta \in G\, h(\beta) = 0.\end{aligned}$$

Hence

.4 $$\text{knap } W \text{ zr } h \in \text{Zr } \overline{h}$$

and we conclude with the aid of .1, .2, .3, .4 and 2.12.3 that

$$\text{Al } \overline{h} \text{ dmn } W\, x\ (0 \leqslant \underline{\mathsf{D}}\ W\, g\, h\, x \leqslant \overline{\mathsf{D}}\ W\, g\, h\, x \leqslant \infty).$$

5.2 LEMMA. If

$$\begin{aligned}&g \in \text{tile } \varphi\ W, \ \ h \in \text{tile } \varphi\ W, \\ &B \in \text{dmn}'\ h, \ \ \overline{h} = \overline{\mathsf{m}}\ h\ \varphi,\end{aligned}$$

then

$$\text{Al } \overline{h}\ (B \text{ dmn } W)\, x\ (0 \leqslant \mathsf{D}\ W\, g\, h\, x < \infty).$$

Proof. Let us agree that

$$R \text{ is the set of positive rational numbers}$$

and

$$\overline{h} = \overline{\mathsf{m}}\ h\ \varphi.$$

We divide the remainder of the proof into three parts.

Part 1. If $r \in R$, $s \in R$, $r < s$ and

$$A = B \cap \mathsf{E}\, x\, (\underline{\mathsf{D}}\, W\, g\, h\, x < r < s < \overline{\mathsf{D}}\, W\, g\, h\, x),$$

then

$$\overline{h}(A) = 0.$$

Proof. Use 4.28 and 4.5.5 to check

$$s \cdot \overline{h}(A) \leqslant \overline{g}(A) \leqslant r \cdot \overline{h}(A) \leqslant r \cdot h(B) < \infty.$$

Accordingly

$$(s - r) \cdot \overline{h}(A) \leqslant 0 \text{ and } \overline{h}(A) = 0.$$

Part 2. If

$$A = B \cap \mathsf{E}\, x\, (\overline{\mathsf{D}}\, W\, g\, h\, x = \infty),$$

then

$$\overline{h}(A) = 0.$$

Proof. Suppose $0 < N < \infty$ and use 4.28.2 to check

$$N \cdot \overline{h}(A) \leqslant \overline{g}(A) \leqslant g(B) < \infty.$$

The arbitrary nature of N assures us that $\overline{h}(A) = 0$.

Part 3. Al $\overline{h}$ $(B \text{ dmn } W)\, x$

$$(0 \leqslant \underline{\mathsf{D}}\ W\ g\ h\ x = \overline{\mathsf{D}}\ W\ g\ h\ x < \infty).$$

Proof. B dmn $W \sim \mathsf{E}\ x\ (0 \leqslant \underline{\mathsf{D}}\ W\ g\ h\ x = \overline{\mathsf{D}}\ W\ g\ h\ x < \infty)$

$$\begin{aligned} &\subset \text{dmn } W \sim \mathsf{E}\ x\ (0 \leqslant \underline{\mathsf{D}}\ W\ g\ h\ x \leqslant \overline{\mathsf{D}}\ W\ g\ h\ x \leqslant \infty) \\ &\quad \cup (B \cap \mathsf{E}\ x\ (\overline{\mathsf{D}}\ W\ g\ h\ x = \infty)) \\ &\quad \cup \bigcup r \in R \bigcup s \in R\ (B \cap \mathsf{E}\ x\ (\underline{\mathsf{D}}\ W\ g\ h\ x < r < s < \overline{\mathsf{D}}\ W\ g\ h\ x)). \end{aligned}$$

Reference to 5.1 and Parts 1 and 2 completes the proof of Part 3.

We complete the proof of the theorem by referring to 3.15 in the light of Part 3.

5.3 THEOREM. If

$$g \in \text{tile } \varphi\ W,\quad h \in \text{tile } \varphi\ W,\quad \overline{h} = \overline{\mathsf{m}}\ h\ \varphi,$$

then

$$\text{Al } \overline{h} \text{ dmn } W\ x\ (0 \leqslant \mathsf{D}\ W\ g\ h\ x < \infty).$$

Proof. Let

$$\begin{aligned} A &= \text{dmn } W \sim \mathsf{E}\ x\ (0 \leqslant \mathsf{D}\ W\ g\ h\ x < \infty), \\ F &= \text{dmn}'\ h. \end{aligned}$$

From 5.2 we see

$$\overline{h}(\beta A) = 0 \text{ whenever } \beta \in F.$$

According to 4.29.1

$$\overline{h}(A) = 0$$

and the proof is complete.

5.4 THEOREM. If

$$g \in \text{localtile}\ \varphi\ W,\ \ h \in \text{localtile}\ \varphi\ W,$$
$$\overline{h} = \overline{\mathsf{m}}\ h\ \varphi,$$

then

$$\text{Al}\ \overline{h}\ \text{dmn}\ W\ x\ (0 \leqslant \mathsf{D}\ W\ g\ h\ x < \infty).$$

Proof. Let

$$F = \text{tiled}\ g\ \varphi\ W \cap \text{tiled}\ h\ \varphi\ W,\ \ A = \text{dmn}\ W,$$
$$g_1 = \text{strc}\ g\ F,\ \ h_1 = \text{strc}\ h\ F,\ \ \overline{h}_1 = \overline{\mathsf{m}}\ h_1\ \varphi.$$

Because of 3.12.3 we know

.1 $$\text{Webalm}\ \varphi\ W\ F$$

and hence because of 4.14 we learn

$$g_1 \in \text{tile}\ \varphi\ W,\ \ h_1 \in \text{tile}\ \varphi\ W.$$

Since 2.12.7 guarantees that

$$\text{Zr}\ \overline{h}_1 \subset \text{Zr}\ \overline{h},$$

we infer from 5.3 that

.2 $$\text{Al}\ \overline{h}\ A\ x\ (0 \leqslant \mathsf{D}\ W\ g_1\ h_1\ x < \infty).$$

Now, because of .1 and 3.17,

.3 $$\text{Al}\ \varphi\ A\ x\ (\mathsf{D}\ W\ g\ h\ x = \mathsf{D}\ W\ g_1\ h_1\ x)$$

and hence, because of 2.12.3,

$$\text{Al}\ \overline{h}\ A\ x\ (\mathsf{D}\ W\ g\ h\ x = \mathsf{D}\ W\ g_1\ h_1\ x).$$

Because of .2 and .3 the desired conclusion is at hand.

5.5 THEOREM. If

$$\varphi \in \text{basetile}\ W,\ \ f \in \text{localtile}\ \varphi\ W,$$

then

$$\text{Al}\ \varphi\ \text{dmn}\ W\ x\ (0 \leqslant \mathsf{D}\ W\ f\ \varphi\ x < \infty).$$

Proof. Let

$$g = f,\ \ h = \varphi,\ \ \overline{h} = \overline{\mathsf{m}}\ h\ \varphi.$$

Since, according to 2.13,

$$\varphi = \overline{h},$$

the desired conclusion follows from 5.4.

5.6 DEFINITION. $\overline{\text{mr}}\ F\ \varphi = \mathsf{E}\ f\ [f = \overline{\mathsf{m}}\ g\ \varphi \text{ for some } g \in F]$.

With the help of 4.20, 4.5.3 and 4.7.9 we easily check

5.7 THEOREM. If $F = \text{tile}\ \varphi\ W$ and $M = \overline{\text{mr}}\ F\ \varphi$, then:

.1 $g \in \text{tile}\ h\ W$ whenever $g \in M$ and $h \in M$;

.2 $M \subset \text{fairtile}\ W$;

.3 if $g \in M$ and $h \in M$, then

$$\text{Al}\ h\ \text{dmn}\ W\ x\ (0 \leqslant \mathsf{D}\ W\ g\ h\ x < \infty).$$

Families M for which the above conclusions hold interest us very much. Other examples will appear presently.

5.8 THEOREM. If

$$f \in \text{tile } \varphi\, W,\ \overline{f} = \overline{\text{m}}\, f\, \varphi,$$
$$\varphi_1 = \text{sct } \varphi\, A,\ F = \text{sct } \overline{f}\, A,\ \overline{f}_1 = \overline{\text{m}}\, f\, \varphi_1,$$

then

$$f \in \text{tile } \varphi_1\, W \quad \text{and} \quad F = \overline{f}_1 \in \text{contile } \varphi_1\, W.$$

Proof. That

$$f \in \text{tile } \varphi_1\, W$$

is a consequence of 4.7.9. That

$$F = \overline{f}_1 \in \text{contile } \varphi_1\, W$$

is now a consequence of 2.15 and 4.20.

5.9 LEMMA. If

$$F \in \text{contile } \varphi\, W,\ f \in \text{tile } \varphi\, W,$$
$$0 < \lambda < \infty,\ B \in \text{dmn } F,\ \overline{f} = \overline{\text{m}}\, f\, \varphi,$$
$$\text{Al } \varphi\, B\, x\, (\underline{\text{D}}\, W\, F\, f\, x \leqslant \lambda),$$

then

$$F(B) \leqslant \lambda \cdot \overline{f}\,(B).$$

Proof. Use 4.5, 4.22, 4.28.1.

5.10 THEOREM. If

$$f \in \text{tile } \varphi\, W,\ \ \overline{f} = \overline{\mathsf{m}}\, f\, \varphi,\ \ F = \text{sct}\, \overline{f}\, A,$$

then

$$\text{Al}\, \overline{f}\, (A \text{ dmn } W)\, x\, (\mathsf{D}\, W\, F\, f\, x = 1).$$

Proof. Let

$$\varphi_1 = \text{sct}\, \varphi\, A,\ \ \overline{f}_1 = \overline{\mathsf{m}}\, f\, \varphi_1 .$$

We learn from 5.8 that

$$f \in \text{tile } \varphi_1\, W,\ \ F = \overline{f}_1 \in \text{contile } \varphi_1\, W.$$

Evidently, because of 5.1, 3.14.3 and 2.12.2,

$$\text{Al}\, \overline{f}_1 \text{ dmn } W\, x\, (0 \leqslant \underline{\mathsf{D}}\, W\, F\, f\, x \leqslant \overline{\mathsf{D}}\, W\, F\, f\, x \leqslant 1).$$

The desired conclusion now follows from 4.29.1 and the

Statement. If $\beta \in \text{dmn}'\, f$, $0 < \lambda < 1$ and

$$B = \beta\, A\ \mathsf{E}\, x\, (\underline{\mathsf{D}}\, W\, F\, f\, x \leqslant \lambda),$$

then

$$\overline{f}\,(B) = 0.$$

Proof. Using 5.9 we infer

$$\overline{f}\,(B) = F(B) \leqslant \lambda \cdot \overline{f}\,(B) \leqslant \lambda \cdot \overline{f}\,(\beta \text{ dmn } W) \leqslant \lambda \cdot f(\beta) < \infty,$$
$$(1 - \lambda) \cdot \overline{f}\,(B) \leqslant 0,\ \ \overline{f}\,(B) \leqslant 0,\ \ \overline{f}\,(B) = 0.$$

5.11 LEMMA. If

$$g \in \text{tile}\ \varphi\ W,\ \ h \in \text{tile}\ \varphi\ W,$$
$$\overline{g} = \overline{\mathrm{m}}\ g\ \varphi,\ \ \overline{h} = \overline{\mathrm{m}}\ h\ \varphi,$$
$$\overline{g}(A) = 0,\ \ \mathsf{D}\ W\ g\ h\ x > 0 \text{ whenever } x \in A,$$

then

$$\overline{h}(A) = 0.$$

5.12 THEOREM. If

$$g \in \text{tile}\ \varphi\ W,\ \ h \in \text{tile}\ \varphi\ W,$$
$$\overline{g} = \overline{\mathrm{m}}\ g\ \varphi,\ \ \overline{h} = \overline{\mathrm{m}}\ h\ \varphi,$$
$$\text{rlm}\ \varphi \in \text{sb}'\ \varphi\ \text{dmn}\ W,$$
$$B = \mathsf{E}\ x\ (\mathsf{D}\ W\ h\ g\ x = 0),$$

then

$$\overline{h}(B) = 0,$$

and a necessary and sufficient condition that

$$\text{Zr}\ \overline{h} \subset \text{Zr}\ \overline{g}$$

is that

$$\overline{g}(B) = 0.$$

Proof. According to 4.30,

$$\overline{h}(B) = 0$$

and the necessity follows. The sufficiency is a consequence of the

Statement. If

.1 $$\overline{g}(B) = 0$$

and

.2 $$A \in \text{Zr}\, \overline{h},$$

then

$$A \in \text{Zr}\, \overline{g}.$$

Proof. Let

.3 $$C = \mathsf{E}\, x\, (\mathsf{D}\, W\, h\, g\, x > 0)$$

and

.4 $$D = \text{rlm}\, \varphi \sim \mathsf{E}\, x\, (0 \leqslant \mathsf{D}\, W\, h\, g\, x < \infty).$$

Because of 2.2.10 and 2.12.3,

.5 $$\text{rlm}\, \varphi \sim \text{dmn}\, W \in \text{Zr}\, \varphi \subset \text{Zr}\, \overline{g}.$$

Because of 5.3,

.6 $$\text{dmn}\, W \sim \mathsf{E}\, x\, (0 \leqslant \mathsf{D}\, W\, h\, g\, x < \infty) \in \text{Zr}\, \overline{g}.$$

Because of .5 and .6, $\overline{g}(D) = 0$ and

.7 $$\overline{g}(AD) = 0.$$

Because of .2, $\overline{h}(A) = 0$ and

.8 $$\overline{h}(AC) = 0.$$

Because of .3, .8 and 5.11,

.9 $$\overline{g}(AC) = 0.$$

Because of .1,

.10 $$\overline{g}(AB) = 0.$$

Because of .7, .9 and .10,

$$A = A \cap \operatorname{rlm} \varphi = AB \cup AC \cup AD \in \operatorname{Zr} \overline{g}.$$

5.13 THEOREM. If

$$g \in \operatorname{tile} \varphi\, W, \quad h \in \operatorname{tile} \varphi\, W,$$
$$\overline{g} = \overline{\mathsf{m}}\, g\, \varphi, \quad \overline{h} = \overline{\mathsf{m}}\, h\, \varphi, \quad G = \operatorname{sct} \overline{g}\, A,$$

then

$$\operatorname{Al} \overline{h}\ (A \operatorname{dmn} W)\, x\ (0 \leqslant \mathsf{D}\, W\, G\, h\, x = \mathsf{D}\, W\, g\, h\, x < \infty).$$

Proof. Let

$$F = \operatorname{dmn} \varphi \cap \operatorname{dmn} g,$$

and check that

.1 $$\operatorname{Webalm} \varphi\, W\, F,$$
.2 $$0 \leqslant G(\beta) \leqslant g(\beta) \text{ whenever } \beta \in F.$$

Let

$$A_1 = \mathsf{E}\, x\ (0 \leqslant \mathsf{D}\ W\ g\ h\ x < \infty \text{ and } \mathsf{D}\ W\ G\ g\ x = 1),$$
$$A_2 = \mathsf{E}\, x\ (\mathsf{D}\ W\ g\ h\ x = 0 = \mathsf{D}\ W\ G\ h\ x),$$
$$A_3 = A\ \mathsf{E}\, x\ (\mathsf{D}\ W\ g\ h\ x > 0) \sim \mathsf{E}\, x\ (\mathsf{D}\ W\ G\ g\ x = 1),$$
$$A_4 = \mathsf{E}\, x\ (\mathsf{D}\ W\ g\ h\ x = 0) \sim \mathsf{E}\, x\ (\mathsf{D}\ W\ G\ h\ x = 0),$$
$$A_5 = \operatorname{dmn} W \sim \mathsf{E}\, x\ (0 \leqslant \mathsf{D}\ W\ g\ h\ x < \infty).$$

According to 3.19,

$$0 \leqslant \mathsf{D}\ W\ G\ h\ x = \mathsf{D}\ W\ g\ h\ x < \infty \text{ whenever } x \in A_1 \cup A_2.$$

According to 5.10 and 5.11,

$$\overline{h}(A_3) = 0.$$

According to .1, .2 and 3.18, $A_4 \subset \operatorname{knap} W \sim F$, $\varphi(A_4) = 0$ and

$$\overline{h}(A_4) = 0.$$

According to 5.3,

$$\overline{h}(A_5) = 0.$$

Since

$$A \operatorname{dmn} W \subset A_1 \cup A_2 \cup A_3 \cup A_4 \cup A_5,$$

the proof is complete.

5.14 THEOREM. If

$$g \in \operatorname{tile} \varphi\ W,\quad h \in \operatorname{tile} \varphi\ W,$$
$$\overline{g} = \overline{\mathsf{m}}\ g\ \varphi,\quad \overline{h} = \overline{\mathsf{m}}\ h\ \varphi,$$
$$G = \operatorname{sct} \overline{g}\ A,\quad H = \operatorname{sct} \overline{h}\ A,$$
$$C = A \cap \operatorname{dmn} W,$$

then:

.1 $$\text{Al}\,\overline{h}\;C\,x\;(\mathsf{D}\;W\,G\,h\,x = \mathsf{D}\;W\,g\,h\,x);$$
.2 $$\text{Al}\,\overline{h}\;C\,x\;(\mathsf{D}\;W\,g\,H\,x = \mathsf{D}\;W\,g\,h\,x);$$
.3 $$\text{Al}\,\overline{h}\;C\,x\;(\mathsf{D}\;W\,G\,H\,x = \mathsf{D}\;W\,g\,h\,x).$$

Proof. We know .1 because of 5.13. From 5.10 and 3.20 we infer

.4 $$\text{Al}\,\overline{h}\;C\,x\;(\mathsf{D}\;W\,h\,H\,x = 1).$$

We now infer .2 from .4, 5.3 and 3.19; we infer .3 from .1, .4, 5.3 and 3.19.

5.15 THEOREM. If

$$\varphi \in \text{fairtile}\;W,\;\; f \in \text{tile}\;\varphi\;W,\;\; \overline{f} = \overline{\mathsf{m}}\;f\;\varphi,$$
$$\varphi_1 = \text{sct}\;\varphi\;A,\;\; F = \text{sct}\;\overline{f}\;A,\;\; C = A \cap \text{dmn}\;W,$$

then:

.1 $$\text{Al}\;\varphi\;C\,x\;(\mathsf{D}\;W\,F\,\varphi\,x = \mathsf{D}\;W\,f\,\varphi\,x);$$
.2 $$\text{Al}\;\varphi\;C\,x\;(\mathsf{D}\;W\,f\,\varphi_1\,x = \mathsf{D}\;W\,f\,\varphi\,x);$$
.3 $$\text{Al}\;\varphi\;C\,x\;(\mathsf{D}\;W\,F\,\varphi_1\,x = \mathsf{D}\;W\,f\,\varphi\,x).$$

From 5.10 we have at once

5.16 THEOREM. If $\varphi \in$ fairtile W and

$$\psi = \text{sct}\;\varphi\;A,$$

then

$$\text{Al}\;\varphi\;(A\;\text{dmn}\;W)\,x\;(\mathsf{D}\;W\;\psi\;\varphi\,x = 1).$$

5.17 THEOREM. If $\varphi \in$ fairtile W, $A \in$ mbl φ, $\theta =$ sct $\varphi \sim A$, then

$$\text{Al}\ \varphi\ (A\ \text{dmn}\ W)\ x\ (\mathsf{D}\ W\ \theta\ \varphi\ x = 0).$$

5.18 THEOREM. If

$$\varphi \in \text{fairtile}\ W,\ \ A \in \text{mbl}\ \varphi,$$
$$\psi = \text{sct}\ \varphi\ A,\ \ \theta = \text{sct}\ \varphi \sim A,$$

then

$$\text{Al}\ \varphi\ (A\ \text{dmn}\ W)\ x\ (\mathsf{D}\ W\ \theta\ \psi\ x = 0).$$

5.19 THEOREM. If

$$h \in \text{tile}\ \varphi\ W,\ \ g \in \text{gauge},\ \ C = \text{dmn}\ W,$$

and

$$\text{Al}\ \varphi\ C\ x\ (\mathsf{D}\ W\ g\ h\ x = 0),$$

then

$$g \in \text{tile}\ \varphi\ W.$$

Proof. The desired conclusion is a consequence of the

Statement. If

$$r > 0,\ \ B \in \text{dmn}\ g,\ \ W' \in \text{cob}\ W\ B,$$

then there is a $G \in$ cwr φ W' for which

$$\Sigma\ \beta \in G\ g(\beta) \leqslant g(B) + r.$$

Proof. Because of 4.6.5,

$$h \in \text{tile}\ \varphi\ W'.$$

Let

$$C' = \text{dmn}\ W', \ \ F = \text{reach}\ W',$$
$$f = \text{strc}\ g\ F, \ \ \overline{f} = \overline{\mathsf{m}}\ f\ \varphi.$$

Helped by 3.8.4, 3.12.10 and 3.17, we infer

$$\text{Al}\ \varphi\ C'\ x\ (0 = \mathsf{D}\ W\ g\ h\ x = \mathsf{D}\ W'\ g\ h\ x = \mathsf{D}\ W'\ f\ h\ x).$$

Because of this and 4.30,

$$\overline{f}\,(C') = 0.$$

Thus according to 2.2.12 there is such a

$$G \in \text{cvr}\ \varphi\ C'\ \text{dmn}\ f \subset \text{cwr}\ \varphi\ W'$$

that

$$\Sigma\ \beta \in G\ g(\beta) = \Sigma\ \beta \in G\ f(\beta) \leqslant r \leqslant g(B) + r.$$

5.20 DEFINITIONS.

.1 $\overline{\mathsf{a}}\ f\ \varphi$ = the function g on dmn f for which

$$g(\beta) = (\overline{\mathsf{m}}\ f\ \varphi)(\beta\ \text{rlm}\ \varphi)$$

whenever $\beta \in$ dmn f.

.2 $\overline{\mathsf{s}}\ f\ \varphi$ = the function h on dmn f such that for each $\beta \in$ dmn f:

if $(\overline{\mathrm{a}}\ f\ \varphi)(\beta) < f(\beta)$ then $h(\beta) = f(\beta) - (\overline{\mathrm{a}}\ f\ \varphi)(\beta)$;
if $(\overline{\mathrm{a}}\ f\ \varphi)(\beta) = f(\beta)$ then $h(\beta) = 0$.

Our conviction that $\infty - \infty$ is not a number, though $\infty - t = \infty$ if $t < \infty$, is reflected in .2 above.

Thus we have the

5.21 THEOREM. If

$$\varphi \in \text{Measure},\ \ f \in \text{gauge},\ \ \overline{f} = \overline{\mathrm{m}}\ f\ \varphi,$$
$$g = \overline{\mathrm{a}}\ f\ \varphi,\ \ h = \overline{\mathrm{s}}\ f\ \varphi,$$

then:

.1 $$g \in \text{gauge},\ \ h \in \text{gauge},\ \ \text{dmn}\ g = \text{dmn}\ h = \text{dmn}\ f;$$

.2 $$g(\beta) = \overline{f}(\beta\ \text{rlm}\ \varphi)$$

whenever $\beta \in \text{dmn}\ f$;

.3 $$h(\beta) = f(\beta) - g(\beta)$$

whenever $\beta \in \text{dmn}\ f$ and $g(\beta) < f(\beta)$;

.4 $$h(\beta) = 0$$

whenever $\beta \in \text{dmn}\ f$ and $g(\beta) = f(\beta)$;

.5 $$f = g + h.$$

With the help of 4.20, 5.15.1 and 5.19 it is not hard to check the following decomposition

5.22 THEOREM. If

$$\varphi \in \text{fairtile } W,\ \ f \in \text{tile } \varphi\ W,$$
$$g = \overline{\mathsf{a}}\ f\ \varphi,\ \ h = \overline{\mathsf{s}}\ f\ \varphi,$$

then:

.1 $g \in \text{contile } \varphi\ W$;

.2 $h \in \text{tile } \varphi\ W$;

.3 $\text{Al } \varphi \text{ dmn } W\ x\ (\mathsf{D}\ W\ g\ \varphi\ x = \mathsf{D}\ W f\ \varphi\ x)$;

.4 $\text{Al } \varphi \text{ dmn } W\ x\ (\mathsf{D}\ W\ h\ \varphi\ x = 0)$;

.5 $f = g + h$.

6. TILES OF VARIOUS SORTS.

In addition to laying the foundation for the integration of the derivative in the next section, we devote considerable space in this section to presenting various realizations of the tiled functions involved in our hypotheses of Sections 4 through 7. The reader who is eager to integrate as soon as possible should read 6.1 through 6.28, then 6.41, then skip to Section 7.

6.1 DEFINITIONS.

.1 dsjn″ = Ε F (F is a countable disjointed family).

.2 $G \subset\subset F$ if and only if every member of G is a subset of some member of F. In other words, $G \subset\subset F$ if and only if G is a refinement of F.

.3 R is separational if and only if R is such a symmetric reflexive relation that

$$(A, B) \in R$$

whenever A, B, A' and B' are sets for which

$$A \in \text{sb } A' \cap \text{dmn } R, \quad B \in \text{sb } B' \cap \text{dmn } R,$$
$$A' \neq B', \quad (A', B') \in R.$$

The prototypal separational relation which we have in mind is the relation of disjointness. The artificial-sounding requirement that R be reflexive is needed after 6.2.3 below to insure dmn $R = M$ in all cases.

.4 G is separated R if and only if

G is countable, R is separational,
$(A, B) \in R$ whenever $A \in G$ and $B \in G$.

.5 R is finished if and only if R is separational, and corresponding to each countable subfamily C of dmn R there is such a separated R family K that

$$K \subset\subset C, \quad \sigma K = \sigma C.$$

.6 $F \cap\cap G = \mathsf{E}\, C\, (C = AB$ for some $A \in F$ and $B \in G)$.

.7 $\text{gauge}_0 = \mathsf{E}\, f \in \text{gauge}\ (0 \in \text{dmn}\, f$ implies $f(0) = 0)$.

6.2 DEFINITIONS.

.1 undertotal $R = \mathsf{E}\, f \in \text{gauge}_0$ [R is separational and

$$f(B) \geqslant \Sigma\, \beta \in G\, f(B\beta)$$

whenever $B \in \text{dmn}\, f$ and G is separated R] .

.2 underaddable $M = \mathsf{E}\, f$ [there is such a finished relation R that

$$\text{dmn}\, R = M \quad \text{and} \quad f \in \text{undertotal}\, R] .$$

.3 underadd $M = \mathsf{E}\, f \in \text{gauge}_0$ [

$$f(B) \geqslant \Sigma\, \beta \in G\, f(B\beta)$$

whenever $B \in \text{dmn}\, f$ and $G \in \text{dsjn}''\ \text{sb}\, M$] .

Thus if

$$R = \mathsf{E}\, \alpha,\beta\ (\alpha \in M,\ \beta \in M,\ \alpha = \beta \text{ or } \alpha\beta = 0),$$

then

$$\text{dmn}\, R = M \quad \text{and} \quad \text{underadd}\, M = \text{undertotal}\, R.$$

.4 subadd $M = \mathsf{E}\, f \in \text{gauge}_0$ [

$$f(B) \geqslant \Sigma\, \beta \in G\, f(\beta)$$

whenever $B \in \text{dmn}\, f$ and $G \in \text{dsjn}''\ \text{sb}\, M\ \text{sb sb}\, B$] .

.5 undersum $M = \mathsf{E}\, f \in \text{gauge}_0$ [corresponding to each $r > 0$, each $B \in \text{dmn}\, f$ and each countable subfamily C of M there is such a countable subfamily K of M that

$$K \subset \subset C,\ \sigma K = \sigma C,\ f(B) + r \geqslant \Sigma\, \beta \in K\, f\,(B\beta)]\,.$$

.6 Subadd = $\mathsf{E}\, f\ (f \in$ subadd dmn $f)$.

One interesting way in which finished relations arise is given by 6.3.1.

6.3 THEOREMS.

.1 If R is separational and if corresponding to each $A \in$ dmn R and each $B \in$ dmn R there is a finite family H for which

$$B \sim A \subset \sigma H \subset B,\ \text{sng}\ A \cup H \text{ is separated } R,$$

then

$$R \text{ is finished}.$$

.2 underadd $M \subset$ subadd M.

.3 If

$$g \in \text{subadd}\ N,\ \ \text{dmn}\ g \cap \cap P \subset N,$$

then

$$g \in \text{underadd}\ P.$$

.4 underaddable $M \subset$ undersum M.

.5 If $f \in$ Measure and $M =$ mbl f, then

$$f \in \text{underadd}\ M \cap \text{subadd}\ M \cap \text{underaddable}\ M \cap \text{undersum}\ M.$$

.6 Measure $\cap$ undersum $M \subset$ underadd M.

.7 If M is the family of all closed intervals of finite positive length and if

$$R = \mathsf{E}\ \alpha,\beta\ (\alpha \in M; \beta \in M; \alpha = \beta \text{ or } \alpha\beta \notin M),$$

then

$$\text{undertotal } R \subset \text{underaddable } M \subset \text{undersum } M.$$

It turns out that undersum M is quite extensive.

.8 If $A \sim B \in M$ whenever $A \in M$ and $B \in M$, then

$$\text{underadd } M \subset \text{undersum } M.$$

.9 If $N \subset M$, then

$$\text{underadd } M \subset \text{underadd } N, \quad \text{subadd } M \subset \text{subadd } N.$$

.10 If

$$M \cap \cap M \subset M \quad \text{and} \quad g = \text{strc } f\, M,$$

then

$$f \in \text{underadd } M \text{ implies } g \in \text{underadd } M,$$
$$f \in \text{undersum } M \text{ implies } g \in \text{undersum } M.$$

.11 If $M \subset N \subset \text{Join}''\, M$, then

$$\text{undersum } M \subset \text{undersum } N.$$

.12 If $M \subset \text{dmn } f$ and $f \in$ Subadd, then

$$f \in \text{subadd } M.$$

.13 If $f \in \text{subadd } M$, then

$$M \subset \text{dmn } f \quad \text{and} \quad \text{strc } f\, M \in \text{Subadd}.$$

.14 If

$$f \in \text{gauge}_0, \quad M = \text{dmn}\, f, \quad N \text{ is a family},$$

g is the function on N for which

$$g(B) = \sup G \in (\text{dsjn}'' \text{ sb } M \text{ sb sb } B)\ \Sigma\ \beta \in G\ f(\beta)$$

whenever $B \in N$, then:

$g \in$ Subadd;
if $M \subset N$ and $f \in$ Subadd, then $f \subset g$;
if $N \cap \cap P \subset N$, then $g \in$ underadd P.

Proof. The proof of the second conclusion is easy and the third is an immediate consequence of the first and 6.3.3. We now turn to the first conclusion.

Since for every B

$$0 \in \text{dsjn}'' \cap \text{sb } M \cap \text{sb sb } B,$$

we are sure that

$$g \in \text{gauge}_0.$$

We suppose then that

$$B \in N \quad \text{and} \quad G \in \text{dsjn}'' \cap \text{sb } N \cap \text{sb sb } B$$

and complete the proof by establishing the

Statement. $g(B) \geqslant \Sigma\ \beta \in G\ g(\beta)$.

Proof. Since

$$g(B) \geqslant g(A) \text{ whenever } B \supset A \in N,$$

we infer

$$g(B) = \infty = \Sigma\, \beta \in G\; g(\beta)$$

in case $g(\beta) = \infty$ for some $\beta \in G$. Thus we suppose

$$g(\beta) < \infty \text{ whenever } \beta \in G, \quad 0 < r < \infty,$$

and select a positive valued function η for which

$$\Sigma\, \beta \in G\; \eta(\beta) \leqslant r.$$

Use the fact that $f(0) = 0$ if $0 \in M$ to select for each $\beta \in G$

$$H(\beta) \in \text{dsjn}'' \cap \text{sb}\, M \cap \text{sb sb}\, \beta$$

so that

$$0 \notin H(\beta) \text{ and } g(\beta) - \eta(\beta) \leqslant \Sigma\, \alpha \in H(\beta)\, f(\alpha).$$

Let

$$K = \bigcup \beta \in G\; H(\beta)$$

and check that

$$K \in \text{dsjn}'' \cap \text{sb}\, M \cap \text{sb sb}\, B.$$

Hence

$$g(B) \geqslant \Sigma\ \alpha \in K\ f(\alpha) = \Sigma\ \beta \in G\ \Sigma\ \alpha \in H(\beta)\ f(\alpha)$$
$$\geqslant \Sigma\ \beta \in G\ (g(\beta) - \eta(\beta)) \geqslant \Sigma\ \beta \in G\ g(\beta) - r.$$

From the arbitrary nature of r we conclude

$$g(B) \geqslant \Sigma\ \beta \in G\ g(\beta).$$

6.4 DEFINITIONS.

.1 tileweb $f\ \varphi$ = E $W \in$ web [$f \in$ gauge, $\varphi \in$ Measure and corresponding to each $r > 0$ and each $B \in$ dmn $f \cap$ sp dmn W there is a $G \in$ cwr $\varphi\ W$ for which

$$\Sigma\ \beta \in G\ f(\beta) \leqslant f(B) + r\,].$$

.2 paveweb $f\ \varphi$ = E $W \in$ web [$f \in$ gauge, $\varphi \in$ Measure and corresponding to each $r > 0$ and each $B \in$ dmn $f \cap$ sp dmn W there is a $G \in$ dsjn$''$ cwr $\varphi\ W$ for which

$$\Sigma\ \beta \in G\ f(\beta) \leqslant f(B) + r\,].$$

6.5 DEFINITION. pave $\varphi\ W$ = E $f \in$ pretile $\varphi\ W$

(coweb $W \subset$ paveweb $f\ \varphi$).

6.6 DEFINITIONS.

.1 subtiled $f\ \varphi\ W$ = E $B \in$ dmn f ($f \in$ gauge, $\varphi \in$ Measure, $W \in$ web and cob $W\ B \subset$ tileweb $f\ \varphi$).

.2 subpaved $f\ \varphi\ W$ = E $B \in$ dmn f ($f \in$ gauge, $\varphi \in$ Measure, $W \in$ web and cob $W\ B \subset$ paveweb $f\ \varphi$).

6.7 DEFINITION. subtile $\varphi\ W$ = E $f \in$ pretile $\varphi\ W$ [

.1 Webalm $\varphi\ W$ subtiled $f\ \varphi\ W$;

there is a family M for which

.2 $f \in$ undersum M

and

.3 Webalm $\varphi\ W\ M$].

6.8 DEFINITION. subpave φ W = E $f \in$ pretile φ W [

.1 Webalm φ W subpaved f φ W;

there is a family M for which

.2 $f \in$ undersum M

and

.3 Webalm φ W M] .

6.9 DEFINITIONS.

.1 coverpaveweb φ = E W (dsjn″ cwr φ $W \neq 0$).

.2 coverpaved φ W = E B ($\varphi \in$ Measure, $W \in$ web and cob W $B \subset$ coverpaveweb φ).

.3 coverpave W = E $\varphi \in$ covertile W

(Webalm φ W coverpaved φ W).

From *Perfect Blankets*[5] we learn that there are many webs W for which coverpave W is a very extensive family. We shall presently find other examples.

6.10 DEFINITIONS.

.1 measuretile φ W = E $f \in$ subtile φ W

($f \in$ Hull mbl f and Webalm φ W mbl f).

.2 measurepave φ W = E $f \in$ subpave φ W

($f \in$ Hull mbl f and Webalm φ W mbl f).

6.11 DEFINITIONS.

.1 fairpave W = E φ ($\varphi \in$ pave φ W).

.2 goodtile W = E φ ($\varphi \in$ measuretile φ W).

.3 goodpave W = E φ ($\varphi \in$ measurepave φ W).

.4 solidtile W = E $\varphi \in$ goodtile W (rlm $\varphi \in$ sb′ φ dmn W).

6.12 DEFINITIONS.

.1 sbmb F = E A ($A \subset B$ for some $B \in F$).

.2 hew $f = \mathsf{E}\, F$ (sbmb $F \cap$ dmn $f \subset F$).

6.13 THEOREMS.

.1 paveweb $f\ \varphi \subset$ tileweb $f\ \varphi$.

.2 If $W \in$ tileweb $f\ \varphi$ and dmn $W \subset B \in$ dmn f, then $W \in$ covertileweb φ.

.3 If $W \in$ paveweb $f\ \varphi$ and dmn $W \subset B \in$ dmn f, then $W \in$ coverpaveweb φ.

.4 subpaved $f\ \varphi\ W \subset$ subtiled $f\ \varphi\ W \subset$ tiled $f\ \varphi\ W \subset$ covertiled $\varphi\ W$.

.5 coverpaveweb $\varphi \subset$ covertileweb φ.

.6 subpaved $f\ \varphi\ W \subset$ coverpaved $\varphi\ W \subset$ covertiled $\varphi\ W$.

.7 solidtile $W \subset$ goodtile $W \subset$ covertile W.

.8 goodpave $W \subset$ coverpave $W \subset$ covertile W.

6.14 THEOREM. If

$$f \in \text{gauge}, \quad \varphi \in \text{Measure}, \quad W \in \text{web},$$

then:

.1 dmn $f \subset$ tiled $f\ \varphi\ W$ if and only if

$$\text{coweb } W \subset \text{tileweb } f\ \varphi;$$

.2 dmn $f \subset$ subtiled $f\ \varphi\ W$ if and only if

$$\text{coweb } W \subset \text{tileweb } f\ \varphi;$$

.3 dmn $f \subset$ subpaved $f\ \varphi\ W$ if and only if

$$\text{coweb } W \subset \text{paveweb } f\ \varphi;$$

.4 if F is any family of sets and

$$\text{coweb } W \subset \text{covertileweb } \varphi,$$

then

$$F \subset \text{covertiled } \varphi \, W;$$

.5 if F is any family of sets and

$$\text{coweb } W \subset \text{coverpaveweb } \varphi,$$

then

$$F \subset \text{coverpaved } \varphi \, W;$$

.6 if Webalm φ W dmn φ, then

$$\varphi \in \text{covertile } W$$

if and only if

$$\text{coweb } W \subset \text{covertileweb } \varphi;$$

.7 if Webalm φ W dmn φ and

$$\text{coweb } W \subset \text{coverpaveweb } \varphi,$$

then

$$\varphi \in \text{coverpave } W.$$

6.15 THEOREMS.

.1 tile φ W = E $f \in$ pretile φ W

$$(\text{coweb } W \subset \text{tileweb } f \, \varphi).$$

.2 pave φ $W \subset$ tile φ W.

We use 6.12 in formulating

6.16 THEOREM. subtiled $f\ \varphi\ W \in$ hew f.

6.17 THEOREM. If $f \in$ subtile $\varphi\ W$ then $f \in$ tile $\varphi\ W$.

Proof. Use 6.7 to secure such a family M that

$$f \in \text{undersum } M \quad \text{and} \quad \text{Webalm } \varphi\ W\ M.$$

Since

$$f \in \text{pretile } \varphi\ W,$$

the desired conclusion is a consequence of 4.3.2, 4.1.4 and the

Statement. If

$$B \in \text{dmn } f, \quad V \in \text{cob } W\ B, \quad r > 0,$$

then there is a $G \in$ cwr $\varphi\ V$ for which

$$\Sigma\ \beta \in G\ f(\beta) \leqslant f(B) + r.$$

Proof. Let

$$2 \cdot r' = r, \quad A = \text{dmn } V, \quad F = \text{subtiled } f\ \varphi\ W$$

and

$$L = M \cap F.$$

Note that

$$\text{Webalm } \varphi \; V \; L.$$

Since

$$\varphi \in \text{covertile } V,$$

we use 4.13 to secure

.1 $$C \in \text{sb } L \cap \text{cwr } \varphi \; V.$$

Evidently

.2 $$C \subset M$$

and

.3 $$C \subset F.$$

Because of .1 and .2 we see that C is a countable subfamily of M and we use 6.2.5 to secure such a countable subfamily K of M that

.4 $$K \subset \subset C,$$

.5 $$\sigma K = \sigma C,$$

and

.6 $$f(B) + r' \geqslant \Sigma \; \beta \in K \; f(B\beta).$$

Now let

.7 $$H = \bigcup \; \beta \in K \text{ sng } (B\beta).$$

In view of .7, .5 and .1,

$$A \sim\sigma H = A \sim(B\sigma K) \subset A \sim(A\sigma K) = A \sim\sigma K = A \sim\sigma C \in \operatorname{Zr} \varphi$$

and hence

.8 $$A \in \operatorname{sb}' \varphi\, \sigma H.$$

Because of .6 and .7

.9 $$f(B) + r' \geqslant \Sigma\, \beta \in K\, f(B\beta) \geqslant \Sigma\, \alpha \in H\, f(\alpha).$$

Consequently

.10 $$H \subset \operatorname{dmn} f.$$

With the help of .4 and .3 we infer that

.11 $$H \subset \subset K \subset \subset C \subset F$$

and we then use .10, .11, 6.16 and 6.12 in checking

$$H \subset F.$$

Because of this and .8 we are now sure

.12 $$H \in \operatorname{cvr} \varphi\, A\ F.$$

Next let p be such a positive valued function that

$$\Sigma\, \beta \in H\, p(\beta) \leqslant r'.$$

Since

$$H \subset F \subset \text{dmn } f$$

and since

$$\text{strc } V\ \beta \in \text{cob } W\ \beta \subset \text{tileweb } f\ \varphi$$

whenever $\beta \in H$, we use 6.4.1 to secure such a function K' on H that for each $\beta \in H$,

$$K'(\beta) \in \text{cwr } \varphi \text{ strc } V\ \beta$$

and

$$\Sigma\ \alpha \in K'(\beta)\ f(\alpha) \leqslant f(\beta) + p(\beta).$$

Now if

$$G = \cup\ \beta \in H\ K'(\beta),$$

then from .12, 4.9 and .9 we conclude

$$G \in \text{cwr } \varphi\ V$$

and

$$\begin{aligned} \Sigma\ \beta \in G\ f(\beta) &\leqslant \Sigma\ \beta \in H\ f(\beta) + \Sigma\ \beta \in H\ p(\beta) \\ &\leqslant f(B) + r' + r' = f(B) + r. \end{aligned}$$

In contrast with 6.17 it is not hard to find $\varphi \in$ Measure and $W \in$ web so that subpave $\varphi\ W$ is not a subset of pave $\varphi\ W$. For example let, in accordance with 2.2.16,

$$S = \text{sng } 0 \cup \text{sng } 1 \cup \text{sng } 2, \quad \varphi = \text{count } S,$$

also let

$$A = \text{sng } 0 \cup \text{sng } 1,\ \ B = \text{sng } 1,\ \ C = \text{sng } 1 \cup \text{sng } 2,$$
$$T = S \sim B,\ \ f = \text{sct } \varphi\ T,\ \ M = \text{sb } S,$$

finally let W be such a function on S that

$$W(0) = \text{sng } (A, A),\ \ W(1) = \text{sng } (B, B),\ \ W(2) = \text{sng } (C, C)$$

and then check that

$$f \in \text{subpave } \varphi\ W \sim \text{pave } \varphi\ W.$$

With the help of 6.15.1 and 6.14.2 we easily check the following partial converse to 6.17.

6.18 THEOREM. If

$$f \in \text{tile } \varphi\ W \cap \text{undersum } M,\ \ \text{Webalm } \varphi\ W\ M,$$

then

$$f \in \text{subtile } \varphi\ W.$$

6.19 THEOREM. If

$$f \in \text{pave } \varphi\ W \cap \text{undersum } M,\ \ \text{Webalm } \varphi\ W\ M,$$

then

$$f \in \text{subpave } \varphi\ W.$$

6.20 THEOREMS.

.1 measurepave $\varphi\ W \subset$ subpave $\varphi\ W \subset$ subtile $\varphi\ W \subset$ tile $\varphi\ W$.

.2 measurepave $\varphi\ W \subset$ measuretile $\varphi\ W \subset$ subtile $\varphi\ W \subset$ tile $\varphi\ W$.

.3 goodpave $W \subset$ goodtile $W \subset$ fairtile $W \subset$ covertile W.

.4 fairpave $W \subset$ fairtile W.

.5 $f \in$ measuretile φ W

if and only if

$$f \in \text{tile } \varphi\ W \cap \text{Hull mbl } f \quad \text{and} \quad \text{Webalm } \varphi\ W \text{ mbl } f.$$

.6 If

$$f \in \text{pave } \varphi\ W \cap \text{Hull mbl } f \quad \text{and} \quad \text{Webalm } \varphi\ W \text{ mbl } f,$$

then

$$f \in \text{measurepave } \varphi\ W.$$

.7 $\varphi \in$ goodtile W

if and only if

$$\varphi \in \text{fairtile } W \cap \text{Hull mbl } \varphi \quad \text{and} \quad \text{Webalm } \varphi\ W \text{ mbl } \varphi.$$

.8 If

$$\varphi \in \text{fairpave } W \cap \text{Hull mbl } \varphi \quad \text{and} \quad \text{Webalm } \varphi\ W \text{ mbl } \varphi,$$

then

$$\varphi \in \text{goodpave } W.$$

.9 If $f \in$ measuretile φ W, then Webalm φ W mbl$'$ f.

.10 If $\varphi \in$ goodtile W, then Webalm φ W mbl$'$ φ.

.11 If $\varphi \in$ solidtile W, then mbl φ = mbl$''$ φ.

6.21 THEOREM. goodpave W = goodtile $W \cap$ coverpave W.

Proof. Because of 6.20.3 and 6.13.8 we are sure that

.1 $$\text{goodpave } W \subset \text{goodtile } W \cap \text{coverpave } W.$$

The desired conclusion is a consequence of .1, 6.11.3, 6.10.2, 6.8, 6.11.2, 6.10.1, 6.7, 6.9.3, 3.12.1, 3.12.3 and the

Statement. If $\varphi \in \text{goodtile } W \cap \text{coverpave } W$, then

$$\text{mbl } \varphi \cap \text{coverpaved } \varphi\ W \subset \text{subpaved } \varphi\ \varphi\ W.$$

Proof. After assuming that

$$\varphi \in \text{goodtile } W \cap \text{coverpave } W,$$

we divide the proof of the Statement into three parts.

Part 1. If

$$B \in \text{mbl } \varphi \cap \text{coverpaved } \varphi\ W,$$
$$W' \in \text{cob } W\ B,\ \ r > 0,$$
$$C \in \text{dmn } \varphi \cap \text{sp dmn } W',\ \ \varphi(C) = \infty,$$

then there is a $G \in \text{dsjn}''\ \text{cwr } \varphi\ W'$ for which

$$\Sigma\ \beta \in G\ \varphi(\beta) \leqslant \varphi(C) + r.$$

Proof. According to 6.9,

$$W' \in \text{coverpaveweb } \varphi$$

and

$$\text{dsjn}''\ \text{cwr } \varphi\ W' \neq 0.$$

Choose $G \in \text{dsjn}'' \text{ cwr } \varphi\ W'$ and note that

$$\Sigma\ \beta \in G\ \varphi(\beta) \leqslant \infty = \varphi(C) + r.$$

Part 2. If

$$B \in \text{mbl } \varphi \cap \text{coverpaved } \varphi\ W,$$
$$W' \in \text{cob } W\ B,\quad 0 < r < \infty,$$
$$C \in \text{dmn } \varphi \cap \text{sp dmn } W',\quad \varphi(C) < \infty,$$

then there is a $G \in \text{dsjn}'' \text{ cwr } \varphi\ W'$ for which

$$\Sigma\ \beta \in G\ \varphi(\beta) \leqslant \varphi(C) + r.$$

Proof. Let

$$\epsilon = r/(\varphi(C) + r),$$

use 6.10.1 and 2.2.15 to choose $C' \in \text{mbl } \varphi \cap \text{sp } C$ so that

$$\varphi(C') = \varphi(C),$$

let

$$F = \mathsf{E}\ \beta\ [\varphi(\beta \sim C') \leqslant \epsilon \cdot \varphi(\beta C')],$$

and let

$$W'' = \text{cwb } W'\ (F \cap \text{mbl } \varphi).$$

According to 6.20.3, 4.6.5 and 5.18,

$$\text{Al } \varphi \text{ dmn } W'\ x\ [\mathsf{D}\ W'\ (\text{sct } \varphi \sim C')(\text{sct } \varphi\ C')\ x = 0],$$

so that in the light of 6.10.1 and 3.14.1,

$$\text{dmn } W' \sim \text{dmn } W'' \in \text{Zr } \varphi.$$

Use 6.9 to choose

$$G \in \text{dsjn}'' \text{ cwr } \varphi\ W''.$$

Then

$$G \in \text{dsjn}'' \text{ cwr } \varphi\ W'$$

and

$$\begin{aligned} &\Sigma\ \beta \in G\ \varphi(\beta) = \Sigma\ \beta \in G\ (\varphi(\beta \sim C') + \varphi(\beta C')) \\ &\leqslant \Sigma\ \beta \in G\ ((1+\epsilon)\cdot\varphi(\beta C')) = (1+\epsilon)\cdot\varphi(C'\ \sigma G) \\ &\leqslant \varphi(C') + \epsilon\cdot\varphi(C') = \varphi(C) + r\cdot\varphi(C)/(\varphi(C)+r) \\ &< \varphi(C) + r. \end{aligned}$$

Part 3. mbl φ ∩ coverpaved φ W ⊂ subpaved φ φ W.

Proof. Use Parts 1 and 2 in the light of 6.6.2 and 6.4.2.

This completes the proof of Part 3, the Statement, and hence the theorem.

Theorem 6.22, which we do not use, was inspired by Theorem 18.1.35 of Hahn and Rosenthal's *Set Functions*(8). In proving 6.22 we use 4.16.2R(1), 4.17.2R, 4.18.2R.

6.22 THEOREM. If

$$\text{Webalm } \varphi\ W \text{ mbl } \varphi \text{ and } \text{cvr } \varphi \text{ dmn } W \text{ dmn}'\ \varphi \neq 0,$$

then a necessary and sufficient condition that

$$\varphi \in \text{fairtile } W$$

is that

.1 $$\text{Al } \varphi\, A\, x\, (\text{D } W\, \psi\, \varphi\, x = 1)$$

whenever

$$A \subset \text{dmn } W \quad \text{and} \quad \psi = \text{sct } \varphi\, A.$$

Proof. Since 5.16 gives the result in one direction immediately, we assume .1 holds and establish in Part 4 below that $\varphi \in$ fairtile W.

Part 1. Webalm φ W dmn$'$ φ.

Proof. In .1 let A = dmn W and apply 3.14.4 with $\lambda = 1/2$.

Part 2. If $C \in$ dmn$'$ φ, $W' \in$ cob W C and $0 < r < \infty$, then there is such a

$$G \in \text{sb mbl } \varphi \cap \text{cwr } \varphi\, W'$$

that

$$\Sigma\, B \in G\, \varphi(B) \leqslant \varphi(C) + r.$$

Proof. Let

$$A = \text{dmn } W'.$$

If $\varphi(C) = 0$, then $G = 0$ satisfies the required conditions. Assume then $\varphi(C) > 0$, let

$$s = (\varphi(C) + r)/\varphi(C),$$
$$M = \text{E } F\, [F \subset \text{mbl } \varphi \cap \text{reach } W' \sim \text{Zr } \varphi,$$

$$\Sigma\ B \in H\ \varphi(B) \leqslant s \cdot \varphi(A\, \sigma H) \text{ for each finite } H \subset F],$$
$$K = \mathsf{E}\ F\ [\varphi(A \sim \sigma F) = 0].$$

We notice that M is capped and complete the proof of Part 2 after establishing the following

Statement. If $F \in M \sim K$, then F is a proper subset of some member of M.

Proof. Notice that $\varphi(A \sim \sigma F) > 0$, $1 < s < \infty$, and use the facts that

$$\text{Al}\ \varphi\ (A \sim \sigma F)\ x\ [\mathsf{D}\ W\ (\text{sct}\ \varphi\ (A \sim \sigma F))\ \varphi\ x = 1]$$

and

$$\text{Webalm}\ \varphi\ W\ \text{mbl}\ \varphi$$

to select such a $D \in \text{mbl}\ \varphi \cap \text{reach}\ W'$ that

$$0 < \varphi(D) \leqslant s \cdot \varphi(D\, A \sim \sigma F).$$

Let

$$F' \doteq F \cup \text{sng}\ D.$$

F is a proper subset of F' because $\varphi(D \sim \sigma F) > 0$. To see that $F' \in M$, suppose that H is a finite subfamily of F and check that

$$\begin{aligned} \varphi(D) + \Sigma\ B \in H\ \varphi(B) &\leqslant s \cdot \varphi(D\, A \sim \sigma F) + s \cdot \varphi(A\ \ \sigma H) \\ &\leqslant s \cdot [\varphi(A\ D \sim \sigma H) + \varphi(A\ \ \sigma H)] \\ &= s \cdot \varphi(A\, (D \cup \sigma H)). \end{aligned}$$

Hence $F' \in M$ and the proof of the Statement is complete.

According to 4.18.2R, $M \cap K \neq 0$. Choose $G \in M \cap K$ and check that $\varphi(A \sim \sigma G) = 0$. If H is a finite subfamily of G, then

$$\Sigma\, B \in H\ \varphi(B) \leqslant s \cdot \varphi(A \quad \sigma H) \leqslant s \cdot \varphi(C) = \varphi(C) + r.$$

Hence

$$\Sigma\, B \in G\ \varphi(B) \leqslant \varphi(C) + r.$$

G is countable because $\varphi(B) > 0$ whenever $B \in G$, and $\varphi(C) + r < \infty$. Thus $G \in$ sb mbl $\varphi \cap$ cwr φ W' and the proof of Part 2 is complete.

Part 3. dmn $\varphi \subset$ tiled φ φ W.

Proof. Suppose

$$C \in \text{dmn } \varphi, \quad W' \in \text{cob } W\ C, \quad r > 0.$$

If $\varphi(C) < \infty$ the conclusion follows from Part 2. Suppose then that

$$\varphi(C) = \infty$$

and use the hypothesis of the theorem to select

$$H \in \text{cvr } \varphi \text{ dmn } W \text{ dmn}'\ \varphi.$$

Use Part 2 to construct such a function K on H that

$$K(\beta) \in \text{sb mbl } \varphi \cap \text{cwr } \varphi \text{ strc } W'\ \beta$$

whenever $\beta \in H$, and let

$$G = \cup\ \beta \in H\ K(\beta).$$

According to 4.8,

$$G \in \text{sb mbl } \varphi \cap \text{cwr } \varphi\ W'.$$

Since it is also true that

$$\Sigma\ B \in G\ \varphi(B) \leqslant \infty = \varphi(C) + r,$$

it follows that

$$C \in \text{tiled}\ \varphi\ \varphi\ W$$

and the proof of Part 3 is complete.

Part 4. $\varphi \in$ fairtile W.

Proof. Refer to Parts 1 and 3, 4.3.4, 4.3.2, 4.3.1, 4.2, 4.5.1.

6.23 THEOREM. If

$$\varphi \in \text{Hull mbl}\ \varphi, \quad W \in \text{web},$$
$$\text{Webalm}\ \varphi\ W\ \text{mbl}\ \varphi,$$
$$\text{cvr}\ \varphi\ \text{dmn}\ W\ \text{dmn}'\ \varphi \neq 0,$$

then a necessary and sufficient condition that

$$\varphi \in \text{goodtile}\ W$$

is that

$$\text{Al}\ \varphi\ A\ x\ (\mathsf{D}\ W\ \psi\ \varphi\ x = 1)$$

whenever

$$A \subset \text{dmn}\ W \quad \text{and} \quad \psi = \text{sct}\ \varphi\ A.$$

Proof. Use 6.22, 6.20.3, 6.11.2, 4.3.4, 6.20.5.

We shall integrate in the sense of Section 4 of *Product Measures*(2). In this connection we agree here as in that paper that

$$0 \cdot y = y \cdot 0 = 0,$$

no matter what y may be.

6.24 DEFINITIONS.

.1 $u(x)$ is φ measurable in x if and only if

$$\text{rlm } \varphi \cap \mathsf{E}\, x\, (u(x) \geqslant \lambda) \in \text{mbl } \varphi$$

whenever $-\infty \leqslant \lambda \leqslant \infty$.

.2 Massable+ $\varphi = \mathsf{E}\, u \in$ gauge (dmn u = rlm φ and $u(x)$ is φ measurable in x).

6.25 DEFINITIONS.

.1 Cr x A is 1 or 0 according as x is or is not a member of A.

.2 Nr x $F = \Sigma\, A \in F$ Cr x A.

6.26 DEFINITION. Gr u φ = the function g on

$$\mathsf{E}\, \beta \in \text{mbl } \varphi\, (-\infty \leqslant \textstyle\int (\text{Cr } x\, \beta \cdot u(x))\, \varphi\, dx \leqslant \infty)$$

for which

$$g(\beta) = \textstyle\int (\text{Cr } x\, \beta \cdot u(x))\, \varphi\, dx$$

whenever $\beta \in$ dmn g.

Rather obvious is the

6.27 THEOREM. If

$$A \in \text{sb}'\, \varphi\, \sigma F\,,$$

then

$$\text{Alm } \varphi\, x\ (\text{Nr } x\ F - \text{Cr } x\ A \geqslant 0).$$

6.28 THEOREM. If

$$u \in \text{Massable+ } \varphi,\ \ g = \text{Gr } u\ \varphi,$$

then:

.1 $\text{dmn } g = \mathsf{E}\ \beta \in \text{mbl } \varphi\ (\beta \sim\text{zr } u \in \text{mbl}''\ \varphi)$;

.2 $g \in \text{underadd dmn } g \subset \text{undersum dmn } g$;

.3 $\Sigma\ \beta \in F\ g(\beta) = \int (\text{Nr } x\ F \cdot u(x))\ \varphi\ \text{d}x$

whenever F is a countable subfamily of dmn g;

.4 $\Sigma\ \beta \in F\ \varphi(\beta) = \int \text{Nr } x\ F\ \varphi\ \text{d}x$

whenever F is a countable subfamily of $\text{mbl}''\ \varphi$.

6.29 THEOREM. If

$$\begin{gathered}\varphi \in \text{Hull mbl } \varphi,\ \ u \in \text{Massable+ } \varphi,\ \ g = \text{Gr } u\ \varphi,\\ 0 < M < \infty,\ \ \alpha = \mathsf{E}\ x\ (u(x) \leqslant M),\\ V \in \text{tileweb } \varphi\ \varphi,\ \ \text{dmn } V \in \text{dmn}'\ \varphi,\\ \text{reach } V \subset \text{mbl}'\ \varphi \cap \text{sb}'\ \varphi\ \alpha,\end{gathered}$$

then

$$V \in \text{tileweb } g\ \varphi.$$

Proof. Let

$$A = \text{dmn } V,\ \ S = \text{rlm } \varphi,$$

and let w be such a function on S that

$$w(x) = u(x) \cdot \mathrm{Cr}\ x\ \alpha \text{ whenever } x \in S.$$

Next let

$$h = \mathrm{Gr}\ w\ \varphi$$

and notice that

$$\alpha \in \mathrm{mbl}\ \varphi, \quad w \in \mathrm{Massable+}\ \varphi,$$

.1 $$w(x) \leqslant M \text{ whenever } x \in S,$$

.2 $$\mathrm{Al}\ \varphi\ \beta\ x\ (w(x) = u(x)) \text{ whenever } \beta \in \mathrm{reach}\ V,$$

.3 $$w(x) \leqslant u(x) \text{ whenever } x \in S.$$

Since

$$V \in \mathrm{tileweb}\ \varphi\ \varphi,$$

the desired conclusion is a consequence of 6.4.1, 6.28 and the

Statement. If

$$r > 0, \quad B \in \mathrm{dmn}\ g \cap \mathrm{sp}\ A, \quad G \in \mathrm{cwr}\ \varphi\ V,$$

and

$$\Sigma\ \beta \in G\ \varphi(\beta) \leqslant \varphi(A) + r/M,$$

then

$$\Sigma\ \beta \in G\ g(\beta) \leqslant g(B) + r.$$

Proof. Since

$$\varphi \in \text{Hull mbl } \varphi$$

we so choose

.4 $$A' \in \text{mbl } \varphi \text{ sp } A \text{ sb } B$$

that

.5 $$\varphi(A') = \varphi(A) < \infty.$$

Since

$$h(A') \leqslant M \cdot \varphi(A') < \infty$$

we are sure

.6 $$h(A') < \infty.$$

Because of .5 and 2.7,

$$\begin{aligned}&\varphi(A' \sim \sigma G) = \varphi(A \sim \sigma G) = 0,\\ &A' \in \text{sb}' \; \varphi \; \sigma G,\end{aligned}$$

and according to 6.27,

.7 $$\text{Alm } \varphi \, x \, (\text{Nr } x \; G - \text{Cr } x \; A' \geqslant 0).$$

From .6, 6.28.3, .7, .1, 6.28.4 and .5 we infer

$$\begin{aligned}-\infty &< \Sigma \, \beta \in G \; h(\beta) - h(A')\\ &= \textstyle\int (\text{Nr } x \; G \cdot w(x)) \, \varphi \, \mathrm{d}x - \int (\text{Cr } x \; A' \cdot w(x)) \, \varphi \, \mathrm{d}x\\ &= \textstyle\int (\text{Nr } x \; G - \text{Cr } x \; A') \cdot w(x) \, \varphi \, \mathrm{d}x\\ &\leqslant M \cdot \textstyle\int (\text{Nr } x \; G - \text{Cr } x \; A') \, \varphi \, \mathrm{d}x\end{aligned}$$

$$
\begin{aligned}
&= M \cdot (\int \mathrm{Nr}\, x\; G\; \varphi\, \mathrm{d}x - \int \mathrm{Cr}\, x\; A'\; \varphi\, \mathrm{d}x) \\
&= M \cdot (\Sigma\, \beta \in G\; \varphi(\beta) - \varphi(A')) \\
&= M \cdot (\Sigma\, \beta \in G\; \varphi(\beta) - \varphi(A)) \\
&\leqslant M \cdot (r/M) = r.
\end{aligned}
$$

Accordingly

.8 $$\Sigma\, \beta \in G\; h(\beta) \leqslant h(A') + r.$$

From .2, .8, .3 and .4 we conclude

$$
\begin{aligned}
\Sigma\, \beta \in G\; g(\beta) = \Sigma\, \beta \in G\; h(\beta) &\leqslant h(A') + r. \\
&\leqslant g(A') + r \leqslant g(B) + r.
\end{aligned}
$$

6.30 THEOREM. If

$$
\begin{aligned}
&\varphi \in \text{Hull mbl}\; \varphi, \quad u \in \text{Massable+}\; \varphi, \quad g = \text{Gr}\; u\; \varphi, \\
&V \in \text{paveweb}\; \varphi\; \varphi, \quad \text{dmn}\; V \in \text{dmn}'\; \varphi, \\
&\alpha \in \text{dmn}'\; g, \quad \text{reach}\; V \subset \text{mbl}'\; \varphi \cap \text{sb}'\; \varphi\; \alpha,
\end{aligned}
$$

then

$$V \in \text{paveweb}\; g\; \varphi.$$

Proof. Let

$$A = \text{dmn}\; V, \quad S = \text{rlm}\; \varphi,$$

and let w be such a function on S that

$$w(x) = u(x) \cdot \text{Cr}\; x\; \alpha \text{ whenever } x \in S.$$

Next let

$$h = \text{Gr } w \ \varphi$$

and notice that

$$\alpha \in \text{mbl } \varphi, \ \ w \in \text{Massable+ } \varphi,$$

.1 $$h(S) < \infty,$$

.2 $$\text{Al } \varphi \ \beta \ x \ (w(x) = u(x)) \text{ whenever } \beta \in \text{reach } V,$$

.3 $$w(x) \leqslant u(x) \text{ whenever } x \in S.$$

The desired conclusion is a consequence of 6.4.2, 6.28, 6.2.3 and the

Statement. If

$$r > 0, \ \ B \in \text{dmn } g \cap \text{sp } A,$$

then there is a $G \in \text{dsjn}'' \text{ cwr } \varphi \ V$ for which

$$\Sigma \ \beta \in G \ g(\beta) \leqslant g(B) + r.$$

Proof. Because of .1 choose $\delta > 0$ so that

.4 $$h(\beta) \leqslant r$$

whenever β is such a member of mbl φ that

$$\varphi(\beta) \leqslant \delta.$$

Since

$$\varphi \in \text{Hull mbl } \varphi$$

we so choose

.5 $$A' \in \text{mbl } \varphi \text{ sp } A \text{ sb } B$$

that

$$\varphi(A') = \varphi(A) < \infty . \tag{.6}$$

Since

$$V \in \text{paveweb } \varphi\ \varphi,$$

we can and do choose such a

$$G \in \text{dsjn}''\ \text{cwr } \varphi\ V \tag{.7}$$

that

$$\Sigma\ \beta \in G\ \varphi(\beta) \leqslant \varphi(A) + \delta .$$

Because of .6 and 2.7,

$$\varphi(A' \sim \sigma G) = \varphi(A \sim \sigma G) = 0.$$

We now infer

$$\begin{aligned}\varphi(\sigma G) \leqslant \Sigma\ \beta \in G\ \varphi(\beta) \leqslant \varphi(A) + \delta &= \varphi(A') + \delta \\ = \varphi(A' \sim \sigma G) + \varphi(A'\ \sigma G) + \delta &= \varphi(A'\ \sigma G) + \delta .\end{aligned}$$

Consequently

$$\varphi(\sigma G \sim A') \leqslant \delta . \tag{.8}$$

We now use .2, .7, .8, .4, .3 and .5 in concluding

$$\begin{aligned}\Sigma\ \beta \in G\ g(\beta) = \Sigma\ \beta \in G\ h(\beta) &= h(\sigma G) \\ &= h(\sigma G\ A') + h(\sigma G \sim A')\end{aligned}$$

$$\leqslant h(A') + h(\sigma G \sim A') \leqslant h(A') + r$$
$$\leqslant g(A') + r \leqslant g(B) + r.$$

In seeking interesting applications of 6.29 and 6.30 we were led to formulate 6.31 and 6.32.

6.31 DEFINITIONS.

.1 Banded $f\ \varphi\ W = \mathsf{E}\ B \in \text{dmn}'\ f$ (for some $\alpha \in \text{dmn}\ \varphi$ and for some M:

$$0 < M < \infty;$$
$$f(\beta) \leqslant M \cdot \varphi(\beta) \text{ whenever } \beta \in \text{dmn}\ f \cap \text{sb}\ \alpha;$$
$$\text{Webalm}\ \varphi\ \text{strc}\ W\ B\ \text{sb}'\ \varphi\ \alpha).$$

.2 Bonded $f\ \varphi\ W = \mathsf{E}\ B \in \text{dmn}'\ f$ (for some $\alpha \in \text{dmn}'\ f$

$$\text{Webalm}\ \varphi\ \text{strc}\ W\ B\ \text{sb}'\ \varphi\ \alpha).$$

.3 bonded $f\ W = \mathsf{E}\ B \in \text{dmn}'\ f$ (for some $\alpha \in \text{dmn}'\ f$

$$\text{Weball strc}\ W\ B\ \text{sb}\ \alpha).$$

6.32 DEFINITIONS.

.1 Bandtile $\varphi\ W = \mathsf{E}\ f$ Webalm $\varphi\ W$ Banded $f\ \varphi\ W$.

.2 Bondtile $\varphi\ W = \mathsf{E}\ f$ Webalm $\varphi\ W$ Bonded $f\ \varphi\ W$.

.3 bondtile $W = \mathsf{E}\ f$ Weball W bonded $f\ W$.

In connection with 6.31 we find 6.33 of interest.

6.33 THEOREMS.

.1 If

$$u \in \text{Massable+}\ \varphi, \quad \text{Webalm}\ \varphi\ W\ \text{mbl}'\ \varphi, \quad g = \text{Gr}\ u\ \varphi,$$

and

$$G = \mathsf{E}\, B \in \text{dmn}'\, g\ [\text{for some } M,\ 0 < M < \infty$$

and

$$\text{Webalm}\ \varphi\ \text{strc}\ W\, B\ \text{sb}'\ \varphi\ \mathsf{E}\, x\ (u(x) \leqslant M)]\,,$$

then

$$G = \text{Banded}\ g\ \varphi\ W.$$

Proof. Let $\alpha = \mathsf{E}\, x\ (u(x) \leqslant M)$ to verify that

$$G \subset \text{Banded}\ g\ \varphi\ W.$$

To obtain the reverse inclusion, suppose

$$B \in \text{dmn}'\, g,\quad \alpha \in \text{dmn}\ \varphi,\quad 0 < M < \infty,$$
$$g(\beta) \leqslant M \cdot \varphi(\beta) \text{ whenever } \beta \in \text{dmn}\ g \cap \text{sb}\ \alpha,$$
$$\text{Webalm}\ \varphi\ \text{strc}\ W\, B\ \text{sb}'\ \varphi\ \alpha.$$

Because of 3.12.3,

$$\text{Webalm}\ \varphi\ \text{strc}\ W\, B\ (\text{mbl}'\ \varphi \cap \text{sb}'\ \varphi\ \alpha)$$

and thus the proof will be complete when we establish

Statement 1. $\text{mbl}'\ \varphi \cap \text{sb}'\ \varphi\ \alpha \subset \text{sb}'\ \varphi\ \mathsf{E}\, x\ (u(x) \leqslant M)$.

Statement 1 is an easy consequence of

Statement 2. If

$$\epsilon > 0,\quad \beta \in \text{mbl}'\ \varphi \cap \text{sb}'\ \varphi\ \alpha,$$
$$\beta' = \beta \cap \alpha \cap \mathsf{E}\, x\ (u(x) \geqslant M + \epsilon),$$

then

$$\varphi(\beta') = 0.$$

Proof. After noticing that

$$\beta' \in \text{dmn } g \cap \text{sb } \alpha \cap \text{mbl}' \varphi,$$

check that

$$(M + \epsilon) \cdot \varphi(\beta') \leqslant g(\beta') \leqslant M \cdot \varphi(\beta'),$$

and conclude that

$$\varphi(\beta') = 0.$$

The proof of 6.33.1 is complete.

.2 If

$$\varphi \in \text{covertile } W, \text{ Webalm } \varphi \; W \text{ sb}' \varphi \; C,$$

then

$$\text{dmn } W \in \text{sb}' \varphi \; C.$$

Proof. Let

$$A = \text{dmn } W, \; F = \text{sb}' \varphi \; C,$$

and use 4.13 to secure

$$H \in \text{sb } F \cap \text{cwr } \varphi \; W.$$

Clearly then

$$
\begin{aligned}
&A \sim \sigma H \in \mathrm{Zr}\ \varphi,\\
&\sigma H \sim C = \cup\ \beta \in H\ (\beta \sim C) \in \mathrm{Zr}\ \varphi,\\
&A \sim C \subset A \sim \sigma H \cup \sigma H \sim C \in \mathrm{Zr}\ \varphi,\\
&A \in \mathrm{sb}'\ \varphi\ C.
\end{aligned}
$$

With the help of 4.10 it is easy to check the

6.34 LEMMAS.

.1 If

$$\mathrm{Webalm}\ \varphi\ W\ F,\quad \mathrm{cwb}\ W\ F \in \mathrm{tileweb}\ f\ \varphi,$$

then

$$W \in \mathrm{tileweb}\ f\ \varphi.$$

.2 If

$$\mathrm{Webalm}\ \varphi\ W\ F,\quad \mathrm{cwb}\ W\ F \in \mathrm{paveweb}\ f\ \varphi,$$

then

$$W \in \mathrm{paveweb}\ f\ \varphi.$$

6.35 THEOREM. If

$$
\begin{aligned}
&\varphi \in \mathrm{goodtile}\ W,\quad u \in \mathrm{Massable}{+}\ \varphi,\\
&g = \mathrm{Gr}\ u\ \varphi,\\
&F = \mathrm{Banded}\ g\ \varphi\ W \cap \mathrm{mbl}'\ \varphi \cap \mathrm{subtiled}\ \varphi\ \varphi\ W,
\end{aligned}
$$

then

$$F \subset \text{subtiled } g\ \varphi\ W \cap \text{dmn}'\ g.$$

Proof. After checking that $F \subset \text{dmn}'\ g$, observe that the desired conclusion is a consequence of 6.28, 6.6.1 and the

Statement. If

$$B \in F,\quad V \in \text{cob } W\ B,$$

then

$$V \in \text{tileweb } g\ \varphi.$$

Proof. Let

$$W' = \text{strc } W\ B$$

and in accordance with 6.33.1 so choose M and α that

$$0 < M < \infty,\quad \alpha = \mathsf{E}\ x\ (u(x) \leqslant M),$$
$$\text{Webalm } \varphi\ W'\ \text{sb}'\ \varphi\ \alpha.$$

Since

$$V \in \text{coweb } W'$$

we know from 3.12.6

$$\text{Webalm } \varphi\ V\ \text{sb}'\ \varphi\ \alpha.$$

Hence because of 6.20.10 and 3.12,

.1 $$\text{Webalm } \varphi\ V\ (\text{mbl}'\ \varphi \cap \text{sb}'\ \varphi\ \alpha).$$

Now let

$$V' = \text{cwb } V \,(\text{mbl}' \varphi \cap \text{sb}' \varphi \alpha).$$

Clearly,

$$V' \in \text{coweb } V \subset \text{cob } W B \subset \text{tileweb } \varphi \varphi,$$
$$\text{dmn } V' \in \text{dmn}' \varphi, \quad \text{reach } V' \subset \text{mbl}' \varphi \cap \text{sb}' \varphi \alpha.$$

According to 6.29,

$$V' \in \text{tileweb } g \varphi$$

and hence, because of .1 and 6.34.1,

$$V \in \text{tileweb } g \varphi.$$

6.36 THEOREM. If

$$\varphi \in \text{goodpave } W, \quad u \in \text{Massable+ } \varphi,$$
$$g = \text{Gr } u \varphi,$$
$$F = \text{Bonded } g \varphi W \cap \text{mbl}' \varphi \cap \text{subpaved } \varphi \varphi W,$$

then

$$F \subset \text{subpaved } g \varphi W \cap \text{dmn}' g.$$

Proof. The desired conclusion is a consequence of 6.28, 6.6.2 and the

Statement. If

$$B \in F, \quad V \in \text{cob } W B,$$

then

$$V \in \text{paveweb } g \ \varphi.$$

Proof. Let

$$W' = \text{strc } W \ B$$

and so choose $\alpha \in \text{dmn}' \ g$ that

$$\text{Webalm } \varphi \ W' \ \text{sb}' \ \varphi \ \alpha.$$

Since

$$V \in \text{coweb } W',$$

we know from 3.12.6

$$\text{Webalm } \varphi \ V \ \text{sb}' \ \varphi \ \alpha.$$

Hence because of 6.20.10 and 3.12,

.1 $$\text{Webalm } \varphi \ V \ (\text{mbl}' \ \varphi \cap \text{sb}' \ \varphi \ \alpha).$$

Now let

$$V' = \text{cwb } V \ (\text{mbl}' \ \varphi \cap \text{sb}' \ \varphi \ \alpha).$$

Clearly

$$V' \in \text{coweb } V \subset \text{cob } W \ B \subset \text{paveweb } \varphi \ \varphi,$$
$$\text{dmn } V' \in \text{dmn}' \ \varphi, \quad \text{reach } V' \subset \text{mbl}' \ \varphi \cap \text{sb}' \ \varphi \ \alpha.$$

According to 6.30,

$$V' \in \text{paveweb } g \ \varphi$$

and hence, because of .1 and 6.34.2,

$$V \in \text{paveweb } g\ \varphi.$$

6.37 THEOREM. If

$$\varphi \in \text{goodtile } W, \quad u \in \text{Massable+ } \varphi,$$
$$g = \text{Gr } u\ \varphi \in \text{Bandtile } \varphi\ W,$$

then

$$g \in \text{subtile } \varphi\ W \cap \text{contile } \varphi\ W.$$

Proof. Use 6.32.1, 6.20.10, 6.11.2, 6.10.1, 6.7.1 and 3.12 to check that

$$\text{Webalm } \varphi\ W\ (\text{Banded } g\ \varphi\ W \cap \text{mbl}'\ \varphi \cap \text{subtiled } \varphi\ \varphi\ W).$$

Use 6.35 to infer from this that

$$\text{Webalm } \varphi\ W\ (\text{subtiled } g\ \varphi\ W \cap \text{dmn}'\ g).$$

Use 6.28.2, 6.2.5, 6.1.7 and 6.13.7 to infer from this that

$$g \in \text{pretile } \varphi\ W.$$

With the help of 6.28 and 6.7 we conclude that

$$g \in \text{subtile } \varphi\ W.$$

With the help of 6.20.1, 4.3.2 and 4.4 we conclude that

$$g \in \text{contile } \varphi\ W.$$

6.38 THEOREM. If

$$\varphi \in \text{goodpave } W, \ \ u \in \text{Massable+ } \varphi,$$
$$g = \text{Gr } u \ \varphi \in \text{Bondtile } \varphi \ W,$$

then

$$g \in \text{subpave } \varphi \ W \cap \text{contile } \varphi \ W.$$

Proof. Helped by 6.20.3, 6.20.10, 6.28.2, 3.12 and 6.36, we infer

$$g \in \text{subpave } \varphi \ W.$$

Now use 6.20.1 and 4.4.

An interesting application of 6.37 arises from

6.39 THEOREM. If

$$\varphi \in \text{goodtile } W, \ \ u \in \text{Massable+ } \varphi,$$
$$0 < M < \infty, \ \ \text{Alm } \varphi \ x \ (u(x) \leqslant M),$$
$$g = \text{Gr } u \ \varphi,$$

then

$$g \in \text{Bandtile } \varphi \ W.$$

Proof. Let

$$\alpha = \mathsf{E} \ x \ (u(x) \leqslant M)$$

and with the help of 6.33.1 and 6.20.10 infer

$$\begin{aligned}&\mathrm{mbl}'\ \varphi \subset \mathrm{dmn}\ \varphi = \mathrm{sb}'\ \varphi\ \alpha,\\ &\mathrm{Webalm}\ \varphi\ W\ \mathrm{sb}'\ \varphi\ \alpha,\ \ \mathrm{Webalm}\ \varphi\ \mathrm{strc}\ W\ B\ \mathrm{sb}'\ \varphi\ \alpha,\\ &\mathrm{mbl}'\ \varphi \subset \mathrm{Banded}\ g\ \varphi\ W,\\ &g \in \mathrm{Bandtile}\ \varphi\ W.\end{aligned}$$

An interesting application of 6.38 arises from

6.40 THEOREM. If

$$\begin{aligned}&\varphi \in \mathrm{goodpave}\ W,\ \ u \in \mathrm{Massable}{+}\ \varphi,\\ &\textstyle\int u(x)\ \varphi\ \mathrm{d}x < \infty,\ \ g = \mathrm{Gr}\ u\ \varphi,\end{aligned}$$

then

$$g \in \mathrm{Bondtile}\ \varphi\ W.$$

Proof. Let

$$\alpha = \mathrm{rlm}\ \varphi$$

and with the help of 6.20.10 infer

$$\begin{aligned}&\mathrm{mbl}'\ \varphi \subset \mathrm{dmn}\ \varphi = \mathrm{sb}'\ \varphi\ \alpha,\\ &\mathrm{Webalm}\ \varphi\ W\ \mathrm{sb}'\ \varphi\ \alpha,\ \ \mathrm{Webalm}\ \varphi\ \mathrm{strc}\ W\ B\ \mathrm{sb}'\ \varphi\ \alpha,\\ &\alpha \in \mathrm{dmn}'\ g,\\ &\mathrm{mbl}'\ \varphi \subset \mathrm{Bonded}\ g\ \varphi\ W,\\ &g \in \mathrm{Bondtile}\ \varphi\ W.\end{aligned}$$

A number of natural questions arising from 6.37 and 6.38 will be answered presently.

6.41 THEOREM. If

$$u \in \mathrm{Massable}{+}\ \varphi,\ \ g = \mathrm{Gr}\ u\ \varphi,\ \ \overline{g} = \overline{\mathrm{m}}\ g\ \varphi,$$

then:

.1 $g \subset \overline{g}$;

.2 $\overline{g}(B) = \inf \beta \in (\text{dmn } g \cap \text{sp } B)\, g(\beta)$

whenever $B \in \text{dmn } \varphi$;

.3 $\text{mbl } \varphi \subset \text{mbl } \overline{g}$;

.4 $\overline{g} \in \text{Hull mbl } \varphi \subset \text{Hull mbl } \overline{g}$;

.5 if $A \subset A' \in \text{mbl}' \varphi$ and $\varphi(A') = \varphi(A)$, then $\overline{g}(A') = \overline{g}(A)$.

6.42 THEOREM. If

$$g \in \text{gauge}, \quad \varphi \in \text{Measure}, \quad \overline{g} = \overline{\text{m}}\, g\, \varphi,$$
$$\overline{g}(B) = \inf \beta \in (\text{dmn } g \cap \text{sp } B)\, g(\beta)$$

whenever $B \in \text{dmn } \varphi$, then:

.1 if $g \in \text{undersum } M$, then $\overline{g} \in \text{undersum } M$;

.2 if $g \in \text{subtile } \varphi\, W$, then $\overline{g} \in \text{subtile } \varphi\, W$;

.3 if $g \in \text{subpave } \varphi\, W$, then $\overline{g} \in \text{subpave } \varphi\, W$.

Proof. The proof of .1 is fairly straightforward. Instead of giving a direct proof of .2 we prefer to indicate a shortcut through tile $\varphi\, W$: use .1, 6.17, 4.20, 6.18. In the case of .3 we are forced to use a more direct approach.

Thus we assume that

.4 $$g \in \text{subpave } \varphi\, W,$$

and complete the proof of .3 by showing in Part 5 below that $\overline{g} \in \text{subpave } \varphi\, W$.

Part 1. There exists such an M that

$$\overline{g} \in \text{pretile } \varphi\, W \cap \text{undersum } M \quad \text{and} \quad \text{Webalm } \varphi\, W\, M.$$

Proof. Use 6.20.1, 6.42.2, 6.7.

Part 2. If

$$B \in \text{dmn } \varphi \cap \text{subpaved } g \; \varphi \; W$$

and

$$W' \in \text{cob } W \; B,$$

then

$$W' \in \text{paveweb } \overline{g} \; \varphi.$$

Proof. Suppose

$$0 < r < \infty \quad \text{and} \quad C \in \text{dmn } \overline{g} \cap \text{sp dmn } W'.$$

Let

$$W'' = \text{cwb } W' \text{ dmn } \varphi.$$

Check that

$$\text{Webalm } \varphi \; W'' \text{ dmn } \varphi$$

and infer from 4.10 that

.5 $$\text{cwr } \varphi \; W'' \subset \text{sb dmn } \varphi \cap \text{cwr } \varphi \; W'.$$

Now use the hypothesis to select $D \in \text{dmn } g$ as follows: if

$$\overline{g}(C) < \infty,$$

choose $D \in \text{sp } C$ so that

$$g(D) \leqslant \overline{g}(C) + r/2;$$

if

$$\overline{g}(C) = \infty,$$

let

$$D = B.$$

Notice that in either case

.6 $$D \in \text{dmn } g \cap \text{sp dmn } W''$$

and

.7 $$g(D) \leqslant \overline{g}(C) + r/2.$$

Since

$$W'' \in \text{cob } W\, B \subset \text{paveweb } g\, \varphi$$

we can use .6 and 6.4.2 to choose

.8 $$G \in \text{dsjn}'' \text{ cwr } \varphi\, W''$$

so that

.9 $$\Sigma\, \beta \in G\; g(\beta) \leqslant g(D) + r/2.$$

Because of .5, .7 and .9 we infer

$$\Sigma\, \beta \in G\; \overline{g}(\beta) \leqslant \Sigma\, \beta \in G\; g(\beta) \leqslant g(D) + r/2 \leqslant \overline{g}(C) + r$$

and

.10 $$\Sigma\, \beta \in G\, \overline{g}(\beta) \leqslant \overline{g}(C) + r.$$

Because of 6.4.2, .8, .5 and .10,

$$W' \in \text{paveweb}\, \overline{g}\, \varphi.$$

Part 3. dmn $\varphi \cap$ subpaved $g\, \varphi\, W \subset$ subpaved $\overline{g}\, \varphi\, W$.

Proof. Use 6.6.2 and Part 2.

Part 4. Webalm $\varphi\, W$ subpaved $\overline{g}\, \varphi\, W$.

Proof. Use Part 3, .4 and 4.2.1.

Part 5. $\overline{g} \in$ subpave $\varphi\, W$.

Proof. Use 6.8, Part 1 and Part 4.

6.43 THEOREM. If

$$u \in \text{Massable+}\ \varphi, \quad g = \text{Gr}\ u\ \varphi, \quad \overline{g} = \overline{\text{m}}\ g\ \varphi,$$

then:

.1 if $g \in$ subtile $\varphi\, W$, then $\overline{g} \in$ measuretile $\varphi\, W$;

.2 if $g \in$ subpave $\varphi\, W$, then $\overline{g} \in$ measurepave $\varphi\, W$.

Proof. Use 6.41 and 6.42.

6.44 DEFINITIONS.

.1 Underweb $f\, \varphi = \mathsf{E}\ W \in$ web [$f \in$ gauge, $\varphi \in$ Measure, and corresponding to each $r > 0$ and each $B \in$ dmn $f \cap$ sp dmn W there is a set C for which

$$f(C) \leqslant f(B) + r, \text{ Webalm } \varphi\ W \text{ sb } C] .$$

.2 underweb f = E $W \in$ web [$f \in$ gauge, and corresponding to each $r > 0$ and each $B \in$ dmn $f \cap$ sp dmn W there is a set C for which

$$f(C) \leqslant f(B) + r, \text{ Weball } W \text{ sb } C] .$$

.3 Undertiled $f\ \varphi\ W$ = E $B \in$ dmn f [

$$f \in \text{gauge}, \quad \varphi \in \text{Measure}, \quad W \in \text{web},$$
$$\text{cob } W\ B \subset \text{Underweb } f\ \varphi] .$$

.4 undertiled $f\ W$ = E $B \in$ dmn f [

$$f \in \text{gauge}, \quad W \in \text{web},$$
$$\text{cob } W\ B \subset \text{underweb } f] .$$

.5 Undertile $\varphi\ W$ = E $f \in$ gauge

$$\text{Webalm } \varphi\ W \text{ Undertiled } f\ \varphi\ W .$$

.6 undertile W = E $f \in$ gauge

$$\text{Weball } W \text{ undertiled } f\ W .$$

.7 Underpave $\varphi\ W$ = E $f \in$ pretile $\varphi\ W$ [

$$\varphi \in \text{coverpave } W, \quad f \in \text{Undertile } \varphi\ W,$$

and there is such a family M that

$$f \in \text{undersum } M \cap \text{subadd } M,$$
$$\text{Webalm } \varphi\ W\ M] .$$

.8 grandpave $W = \mathsf{E}\, f \in$ coverpave $W \cap$ undertile W [

$$f \in \text{Hull mbl } f \quad \text{and} \quad \text{Weball } W \text{ mbl}' f] .$$

.9 greatpave W = Msr dmn $W \cap$ grandpave W.

It turns out that the theory of grandpave W is very pleasant and that of greatpave W most satisfactory.

6.45 THEOREM. If $\varphi \in$ Measure, then:

.1 underweb $f \subset$ Underweb $f\, \varphi$,

.2 undertiled $f\, W \subset$ Undertiled $f\, \varphi\, W$,

.3 undertile $W \subset$ Undertile $\varphi\, W$.

In line with 6.14 is

6.46 THEOREM. If

$$f \in \text{gauge}, \quad \varphi \in \text{Measure}, \quad W \in \text{web},$$

then:

.1 dmn $f \subset$ Undertiled $f\, \varphi\, W$ if and only if

$$\text{coweb } W \subset \text{Underweb } f\, \varphi;$$

.2 if Webalm $\varphi\, W$ dmn f and

$$\text{coweb } W \subset \text{Underweb } f\, \varphi,$$

then

$$f \in \text{Undertile } \varphi\, W.$$

6.47 THEOREM. If

$$f \in \text{gauge}, \quad W \in \text{web},$$

then:

.1 dmn $f \subset$ undertiled $f\ W$ if and only if

$$\text{coweb } W \subset \text{underweb } f;$$

.2 if Weball W dmn f and

$$\text{coweb } W \subset \text{underweb } f,$$

then

$$f \in \text{undertile } W.$$

6.48 LEMMA. If

$$f \in \text{subadd } M, \quad \text{Webalm } \varphi\ W\ M,$$

then

$$\text{coverpaved } \varphi\ W \cap \text{Undertiled } f\ \varphi\ W \subset \text{subpaved } f\ \varphi\ W.$$

Proof. The desired conclusion is a consequence of 6.6.2 and the

Statement 1. If

$$A \in \text{coverpaved } \varphi\ W \cap \text{Undertiled } f\ \varphi\ W,$$
$$V \in \text{cob } W\ A,$$

then

$$V \in \text{paveweb } f \ \varphi.$$

The conclusion now desired is a consequence of 6.4.2 and the

Statement 2. If

$$r > 0, \ \ B \in \text{dmn } f \cap \text{sp dmn } W,$$

then there is a $G \in \text{dsjn}'' \text{ cwr } \varphi \ V$ for which

$$\Sigma \ \beta \in G \ f(\beta) \leqslant f(B) + r.$$

Proof. Since according to 6.44.3

$$V \in \text{cob } W \ A \subset \text{Underweb } f \ \varphi,$$

we so choose C in accordance with 6.44.1 that

$$f(C) \leqslant f(B) + r, \ \ \text{Webalm } \varphi \ V \text{ sb } C.$$

Let

$$F = \text{sb } C, \ \ V' = \text{cwb } V \ (F \cap M).$$

Because of 3.12,

$$\text{Webalm } \varphi \ V \ (F \cap M).$$

Since according to 6.9.2

$$V' \in \text{cob } W \ A \subset \text{coverpaveweb } \varphi,$$

we can use 6.9.1 and 4.10 so to select

$$G \in \text{dsjn}'' \text{ sb } (F \cap M) \cap \text{cwr } \varphi\ V$$

that, because of 6.2.4,

$$\Sigma\ \beta \in G\ f(\beta) \leqslant f(C) \leqslant f(B) + r.$$

6.49 THEOREM. If

$$\begin{aligned}&\text{Webalm } \varphi\ W \text{ dmn}'\ f,\ \ \text{Webalm } \varphi\ W \text{ dmn } \varphi,\\ &f \in \text{subadd } M,\ \ \text{Webalm } \varphi\ W\ M,\\ &\text{coweb } W \subset \text{coverpaveweb } \varphi \cap \text{Underweb } f\ \varphi,\end{aligned}$$

then

$$f \in \text{pave } \varphi\ W.$$

Proof. Because of 6.14.7 and 6.9.3,

$$\varphi \in \text{coverpave } W \subset \text{covertile } W.$$

Consequently

$$f \in \text{pretile } \varphi\ W.$$

Because of 6.14.5, 6.46.1 and 6.48,

$$\begin{aligned}\text{dmn } f &\subset \text{coverpaved } \varphi\ W \cap \text{Undertiled } f\ \varphi\ W\\ &\subset \text{subpaved } f\ \varphi\ W.\end{aligned}$$

According to this and 6.14.3,

$$\text{coweb } W \subset \text{paveweb } f\ \varphi$$

and the desired conclusion follows from 6.5.

6.50 THEOREM. If

$$\text{Webalm}\ \varphi\ W\ \text{dmn}'\ f,\quad \text{Webalm}\ \varphi\ W\ \text{dmn}\ \varphi,$$
$$f \in \text{subadd}\ M \cap \text{undersum}\ M,\quad \text{Webalm}\ \varphi\ W\ M,$$
$$\text{coweb}\ W \subset \text{coverpaveweb}\ \varphi \cap \text{Underweb}\ f\ \varphi,$$

then

$$f \in \text{Underpave}\ \varphi\ W \cap \text{pave}\ \varphi\ W.$$

Proof. Because of 6.49,

$$f \in \text{pave}\ \varphi\ W \subset \text{pretile}\ \varphi\ W.$$

Because of 6.14.7 and 6.46.2,

$$\varphi \in \text{coverpave}\ W \quad \text{and} \quad f \in \text{Undertile}\ \varphi\ W.$$

Because of 6.44.7,

$$f \in \text{Underpave}\ \varphi\ W.$$

From 6.48 and 6.3.2 we infer almost at once the

6.51 THEOREM. Underpave φ $W \subset$ subpave φ W.

By taking M = mbl f it is now easy, with the help of 6.45, 6.3.5 and 6.44, to check

6.52 THEOREM. If

$$f \in \text{grandpave}\ W,\quad \varphi \in \text{grandpave}\ W,$$

then

$$f \in \text{Underpave } \varphi \ W.$$

Of great interest to us is the

6.53 THEOREM. If

$$f \in \text{grandpave } W, \quad g \in \text{grandpave } W,$$

then

$$f \in \text{measurepave } g \ W.$$

Proof. Use 6.52, 6.51 and 6.10.2.

With the help of 6.44, 6.53, 6.13 and 6.20 it is now easy to check the

6.54 THEOREMS.

.1 greatpave $W \subset$ grandpave $W \subset$ goodpave W.

.2 greatpave $W \subset$ solidtile $W \subset$ goodtile W.

Quite obvious is

6.55 THEOREM. sb $(A \cap B) =$ sb $A \cap$ sb B.

6.56 THEOREM. If

$$\begin{aligned}
&\varphi \in \text{covertile } W \cap \text{Hull mbl } \varphi, \ u \in \text{Massable+ } \varphi,\\
&g = \text{Gr } u \ \varphi, \ \overline{g} = \overline{\text{m}} \ g \ \varphi,\\
&A \in \text{bonded } g \ W \cap \text{dmn}' \ \varphi,
\end{aligned}$$

then

$$\text{cob } W \ A \cap \text{underweb } \varphi \subset \text{underweb } \overline{g}.$$

Proof. Let

$$S = \text{rlm } \varphi, \quad W' = \text{strc } W\, A$$

and in accordance with 6.31.3 so choose

$$\alpha \in \text{dmn}'\, g$$

that

.1 $$\text{Weball } W' \text{ sb } \alpha.$$

Let w be such a function on S that

$$w(x) = u(x) \cdot \text{Cr } x\, \alpha \quad \text{whenever } x \in S,$$

and let

$$h = \text{Gr } w\, \varphi, \quad \overline{h} = \overline{\text{m}}\, h\, \varphi.$$

Next notice that

$$\alpha \in \text{mbl } \varphi, \quad w \in \text{Massable+ } \varphi,$$

.2 $$h(S) < \infty,$$

$$u(x) = w(x) \quad \text{whenever } x \in \alpha,$$

.3 $$\overline{h}(\beta) = \overline{g}(\beta) \quad \text{whenever } \beta \in \text{sb } \alpha.$$

The desired conclusion is a consequence of 6.44.2 and the

Statement. If

$$V \in \text{cob } W\, A \cap \text{underweb } \varphi, \quad r > 0,$$

$$B \in \text{dmn } \overline{g} \cap \text{sp dmn } V,$$

then there is a set C for which

$$\overline{g}(C) \leqslant \overline{g}(B) + r, \text{ Weball } V \text{ sb } C.$$

Proof. Because of .2 choose $\delta > 0$ so that

.4 $$h(\beta) \leqslant r \text{ whenever } \beta \in \text{mbl } \varphi \text{ and } \varphi(\beta) \leqslant \delta.$$

Let

$$D = \text{dmn } V$$

and note

.5 $$\text{dmn } V = D \subset B.$$

Because of .5 and the fact that

$$V \in \text{underweb } \varphi,$$

we can and do choose such a set K that

.6 $$\varphi(K) \leqslant \varphi(D) + \delta$$

and

.7 $$\text{Weball } V \text{ sb } K.$$

Now let

.8 $$C = K \cap \alpha.$$

Since evidently

$$V \in \text{coweb } W',$$

we infer from 3.12, .1, .7, .8 and 6.55 that

.9 $$\text{Weball } V \text{ sb } C.$$

Because of .9 and 6.33, it is easy to make sure that

$$D \in \text{sb}' \varphi\ C.$$

Accordingly,

$$\varphi(D) \leqslant \varphi(DC) + \varphi(D \sim C) = \varphi(DC) \leqslant \varphi(D)$$

and

.10 $$\varphi(D) = \varphi(DC).$$

Now choose

.11 $$D' \in \text{mbl } \varphi \cap \text{sp } D \quad \text{and} \quad C' \in \text{mbl } \varphi \cap \text{sp } C \cap \text{sb } \alpha$$

so that

.12 $$\varphi(D) = \varphi(D'), \ \ \varphi(C) = \varphi(C').$$

Because of .12, .11, .10, .8 and .6,

$$\begin{aligned} \varphi(C' \sim D') &= \varphi(C') - \varphi(D'C') \\ &\leqslant \varphi(C') - \varphi(DC) \\ &= \varphi(C) - \varphi(D) \\ &\leqslant \varphi(K) - \varphi(D) \leqslant \delta. \end{aligned}$$

Consequently, because of .11, .3, 6.41.1 and .4,

$$\overline{g}(C' \sim D') = \overline{h}(C' \sim D') = h(C' \sim D') \leqslant r.$$

With the help of .11, .12, 6.41.5 and .5, we now infer

$$\begin{aligned}\overline{g}(C) \leqslant \overline{g}(C') &\leqslant \overline{g}(C'D') + \overline{g}(C' \sim D') \\ &\leqslant \overline{g}(D') + r = \overline{g}(D) + r \leqslant \overline{g}(B) + r.\end{aligned}$$

Because of this and .9, the proof is complete.

6.57 THEOREM. If

$$\begin{aligned}&\varphi \in \text{covertile } W \cap \text{Hull mbl } \varphi, \ \ u \in \text{Massable+ } \varphi, \\ &g = \text{Gr } u\ \varphi, \ \ \overline{g} = \overline{\text{m}}\ g\ \varphi, \\ &F = \text{bonded } g\ W \cap \text{mbl}'\ \varphi \cap \text{undertiled } \varphi\ W,\end{aligned}$$

then

$$F \subset \text{undertiled } \overline{g}\ W \cap \text{mbl}'\ \overline{g}.$$

Proof. If

$$A \in F,$$

then: according to 6.44.4 and 6.56,

$$\text{cob } W\ A = \text{cob } W\ A \cap \text{underweb } \varphi \subset \text{underweb } \overline{g}$$

and hence

$$A \in \text{undertiled } \overline{g}\ W;$$

according to 6.31.3 and 6.41,

$$A \in \text{mbl}'\ \overline{g}.$$

Of considerable interest to us is

6.58 THEOREM. If

$$\varphi \in \text{grandpave } W, \quad u \in \text{Massable+ } \varphi,$$
$$g = \text{Gr } u\ \varphi \in \text{bondtile } W, \quad \overline{g} = \overline{\text{m}}\ g\ \varphi,$$

then

$$\overline{g} \in \text{grandpave } W.$$

Proof. Because of 6.41.4,

.1 $$\overline{g} \in \text{Hull mbl } \overline{g},$$

and since

$$\text{Zr } \varphi \subset \text{Zr } \overline{g},$$

it follows from 4.7, 6.13.8, 6.54.1 and 6.9 that

$$\overline{g} \in \text{covertile } W,$$
$$\text{coverpaveweb } \varphi \subset \text{coverpaveweb } \overline{g},$$
$$\text{coverpaved } \varphi\ W \subset \text{coverpaved } \overline{g}\ W,$$

.2 $$\overline{g} \in \text{coverpave } W.$$

Because of 3.12, 6.32.3, 6.44.6, 6.44.8 and 6.57,

.3 $$\text{Weball } W \text{ undertiled } \overline{g}\ W,$$

.4 $$\text{Weball } W \text{ mbl}'\ \overline{g}.$$

From .3 and 6.44.6 we infer

.5 $$\bar{g} \in \text{undertile } W.$$

From .2, .5, .1, .4 and 6.44.8 we conclude

$$\bar{g} \in \text{grandpave } W.$$

An interesting application of 6.58 arises from

6.59 THEOREM. If

$$\varphi \in \text{grandpave } W, \ u \in \text{Massable+ } \varphi,$$
$$\textstyle\int u(x)\,\varphi\,\mathrm{d}x < \infty, \ g = \text{Gr } u\ \varphi,$$

then

$$g \in \text{bondtile } W.$$

Proof. Let

$$\alpha = \text{rlm } \varphi$$

and with the help of 6.44.8, 6.31.3 and 6.28.1 infer

$$\text{mbl}'\ \varphi \subset \text{dmn } \varphi = \text{sb } \alpha,$$
$$\text{Weball } W \text{ sb } \alpha, \ \text{Weball strc } W\ B \text{ sb } \alpha,$$
$$\alpha \in \text{dmn}'\ g,$$
$$\text{mbl}'\ \varphi \subset \text{mbl } \varphi = \text{dmn } g = \text{dmn}'\ g = \text{bonded } g\ W,$$
$$g \in \text{bondtile } W.$$

The next two theorems bring out the differential properties of grandpave W and greatpave W.

6.60 THEOREM. If

$$g \in \text{grandpave } W, \quad h \in \text{grandpave } W,$$

then

$$\text{Al } h \text{ dmn } W\, x\, (0 \leqslant \mathsf{D}\ W\, g\, h\, x < \infty).$$

Proof. Use 6.53, 6.20.1, 4.5.3 and 4.3.5 in checking

.1 $g \in \text{localtile } h\ W,$

$h \in \text{localtile } h\ W,$

.2 $h \in \text{basetile } W.$

Now infer the desired conclusion from .2, .1 and 5.5.

From 6.60 and 6.44.9 we have at once

6.61 THEOREM. If

$$g \in \text{greatpave } W, \quad h \in \text{greatpave } W,$$

then

$$\text{Alm } h\, x\, (0 \leqslant \mathsf{D}\ W\, g\, h\, x < \infty).$$

6.62 THEOREM. If

$$g \in \text{greatpave } W, \quad h \in \text{greatpave } W,$$
$$B = \mathsf{E}\, x\, (\mathsf{D}\ W\, h\, g\, x = 0),$$

then

$$h\,(B) = 0$$

and a necessary and sufficient condition that

$$\mathrm{Zr}\ h \subset \mathrm{Zr}\ g$$

is that

$$g(B) = 0.$$

Proof. Let

$$G = \text{greatpave } W.$$

With the help of 6.53 and 6.20.1 we see that

$$\mu \in \text{tile } \nu\ W$$

whenever

$$\mu \in G \quad \text{and} \quad \nu \in G.$$

Also

$$\text{rlm } \mu = \text{dmn } W$$

whenever $\mu \in G$.

If

$$\varphi = h,\ \overline{g} = \overline{\mathrm{m}}\ g\ \varphi,\ \overline{h} = \overline{\mathrm{m}}\ h\ \varphi,$$

then because of 5.12,

$$h(B) = \overline{h}(B) = 0.$$

The necessity now follows.

If

$$g(B) = 0, \quad \varphi = g, \quad \overline{g} = \overline{\mathrm{m}}\, g\, \varphi, \quad \overline{h} = \overline{\mathrm{m}}\, h\, \varphi,$$

then

$$\overline{g}(B) = 0$$

and because of 5.12,

$$\mathrm{Zr}\, h \subset \mathrm{Zr}\, \overline{h} \subset \mathrm{Zr}\, \overline{g} = \mathrm{Zr}\, g.$$

A rather useful topology arises from a web in a natural way.

6.63 DEFINITIONS.

.1 tplg $W = \mathsf{E}\, A$ ($W \in$ web and

$$\text{Weball strc } W\, A \text{ sb } A).$$

.2 tpg W = sb σ reach $W \cap$ tplg W.

.3 topology = $\mathsf{E}\, T$ (T is closed under finite non-empty intersections and arbitrary unions).

Because of its large size, tplg W is never a topology. However, because of 6.64.1 and 6.64.2, it doesn't miss it by much. On the other hand, tpg W in 6.64.3 is completely manageable.

6.64 THEOREMS.

.1 If

$$A \in \text{tplg } W, \quad B \in \text{tplg } W,$$

then

$$A \cap B \in \text{tplg } W.$$

.2 If

$$W \in \text{web}, \quad F \in \text{sb tplg } W,$$

then

$$\sigma F \in \text{tplg } W.$$

.3 If

$$W \in \text{web}, \quad T = \text{tpg } W,$$

then

$$T \in \text{topology}, \quad \sigma T = \sigma \text{ reach } W.$$

6.65 DEFINITIONS.

.1 net = E F ($\sigma G \in G$ whenever G is such a non-vacuous subfamily of F that $\pi G \neq 0$).

.2 net′ = E $F \in$ net (every disjointed subfamily of F is countable).

.3 net″ = E $F \in$ net′ ($\sigma F \sim \beta \in$ Join″ F whenever $\beta \in F$).

6.66 DEFINITION. top F = E $B \in F$ (for each A, $B \subset A \in F$ implies $B = A$).

From 2.7C[(9)], 2.12C and 2.6C we have

6.67 THEOREM. If $F \in$ net, then:

.1 if $G \subset F$, then top G is such a disjointed subfamily of G that

$$G \subset\subset \text{top } G, \quad \sigma \text{ top } G = \sigma G;$$

.2 if $A \in F$, $B \in F$, $A B \neq 0$, then $A B = A$ or $A B = B$.

6.68 THEOREM. If $F \in$ net, then

$$\text{underadd } F \subset \text{undersum } F.$$

Proof. Use 6.2.3, 6.2.5 and 6.67.1, with $K =$ top C.

6.69 DEFINITION. netweb $F =$ the function W on σF for which

$$W(x) = \mathsf{E}\, \alpha, \beta\ (\alpha \in F, \beta \in F \text{ and } x \in \beta \subset \alpha)$$

whenever $x \in \sigma F$.

6.70 THEOREM. If

$$F \in \text{net}, \quad W = \text{netweb } F, \quad T = \text{tpg } W,$$

then

$$W \in \text{web}, \quad \text{dmn } W = \sigma F, \quad \text{reach } W = F \sim \text{sng } 0 \subset F,$$
$$F \subset T \in \text{topology}.$$

6.71 DEFINITION. netsum $F = \mathsf{E}\, f$ (dmn $f = F \in$ net$'$,

$$f \in \text{underadd } F, \ \text{ Weball netweb } F \text{ dmn}' f).$$

6.72 THEOREM. If

$$\text{netsum } F \neq 0,$$

then

$$F \in \text{net}' \text{ and } F \cap \cap F \subset F.$$

6.73 THEOREM. If

$$F \in \text{net}', \quad W = \text{netweb } F, \quad \psi \in \text{Msr } \sigma F,$$

then

$$\text{coweb } W \subset \text{coverpaveweb } \psi.$$

Proof. Suppose $W' \in \text{coweb } W$ and let

$$F' = \text{reach } W', \quad G = \text{top } F'.$$

Use 6.67.1 and 6.65.2 to conclude that

$$G \in \text{dsjn}'' \text{ cwr } \psi \ W' \neq 0$$

and

$$W' \in \text{coverpaveweb } \psi.$$

Helped by 6.46 we next check

6.74 THEOREM. If

$$f \in \text{netsum } F, \quad W = \text{netweb } F, \quad \varphi = \text{count } \sigma F,$$

then

$$\text{coweb } W \subset \text{Underweb } f \ \varphi.$$

Theorem 6.75 below answers questions asked us in 1954 by S. Kakutani.

6.75 THEOREM. If

$$W = \text{netweb } F, \ \ \varphi = \text{count } \sigma F,$$

then:

.1 netsum $F \subset$ Underpave $\varphi\ W \cap$ pave $\varphi\ W$;

.2 if

$$g \in \text{netsum } F, \ \ h \in \text{netsum } F, \ \ \overline{h} = \overline{\mathsf{m}}\ \varphi\ h,$$

then

$$\text{Alm } \overline{h}\ x\ (0 \leqslant \mathsf{D}\ W\ g\ h\ x < \infty).$$

Proof. We see that .1 is a consequence of 6.73, 6.74, 6.68, 6.3.2 and 6.50.

Now that we know .1, we can verify .2 either with the help of 6.15.2 and 5.3, or alternatively with the help of 6.51, 6.20.1 and 5.3.

6.76 DEFINITION. $\text{sng}''\ \beta = \text{sng}\ \beta \cap \mathsf{E}\ t\ (\beta \neq 0)$.

Thus $\text{sng}''\ \beta = \text{sng}\ \beta$ unless $\beta = 0$; $\text{sng}''\ 0 = 0$.

6.77 THEOREM. If

$$F \in \text{net}'', \ \ f \in \text{netsum } F,$$
$$\varphi = \text{count } \sigma F, \ \ \overline{f} = \overline{\mathsf{m}}\ f\ \varphi,$$

then

$$F \subset \text{mbl } \overline{f}.$$

Proof. Let $S = \sigma F$ and divide the remainder of the proof into two parts.

Part 1. If $B \in F$ and $A \in F$, then

$$f(B) \geqslant \overline{f}(BA) + \overline{f}(B \sim A).$$

Proof. First choose a countable subset H of F so that

$$S \sim A = \sigma H.$$

Let

$$H' = \text{top } H, \ \ G' = \text{U } \beta \in H' \text{ sng}'' (B\beta),$$
$$G'' = \text{sng}'' (BA), \ \ G = G'' \cup G'.$$

Since

$$\sigma \text{ sng}'' \alpha = \alpha \text{ whenever } \alpha \subset S,$$

it is clear that

$$G \in \text{dsjn}'' \text{ sb } F,$$
$$B \sim A = \sigma G', \ \ BA = \sigma G'',$$
$$\sigma G = B, \ \ G' \cap G'' = 0.$$

Hence

$$f(B) \geqslant \Sigma\ \beta \in G\ f(\beta) = \Sigma\ \beta \in G''\ f(\beta) + \Sigma\ \beta \in G'\ f(\beta)$$
$$\geqslant \overline{f}(BA) + \overline{f}(B \sim A).$$

Part 2. $F \subset \text{mbl}\, \overline{f}$.

Proof. Suppose $A \in F$ and use Part 1 and 2.14 with $C = S$ to check that $A \in \text{mbl}\, \overline{f}$.

6.78 THEOREM. If

$$f \in \text{netsum } F, \ \ W = \text{netweb } F,$$
$$\varphi = \text{count } \sigma F, \ \ \overline{f} = \overline{\text{m}}\, f\, \varphi,$$

then

$$\text{coweb } W \subset \text{underweb } \overline{f} \subset \text{Underweb } \overline{f}\ \varphi .$$

Proof. If

$$W' \in \text{coweb } W,\ \ r > 0,\ \ B \in \text{dmn } \overline{f} \cap \text{sp dmn } W',$$

then with the help of 6.70 and 2.2.12 we can so find

$$C \in \text{tpg } F \text{ sp } B$$

that

$$\overline{f}(C) \leqslant \overline{f}(B) + r,\ \ \text{Weball } W' \text{ sb } C.$$

Consequently

$$\text{coweb } W \subset \text{underweb } \overline{f},$$

and reference to 6.45.1 completes the proof.

6.79 THEOREM. If

$$F \in \text{net}'',\ f \in \text{netsum } F,\ \ W = \text{netweb } F,$$
$$\varphi = \text{count } \sigma F,\ \ \overline{f} = \overline{\text{m}}\ f\ \varphi,$$

then

$$\overline{f} \in \text{greatpave } W \cap \text{pave } \varphi\ W.$$

Proof. From 6.73 and 6.14.7 we infer

.1 $$\overline{f} \in \text{coverpave } W$$

and from 6.78 and 6.47.2 we infer

.2 $$\overline{f} \in \text{undertile } W.$$

Next we use 6.77 and 6.71 in seeing that

.3 $$\overline{f} \in \text{Hull mbl } \overline{f}$$

and

.4 $$\text{Weball } W \text{ mbl}' \, \overline{f}.$$

From .1, .2, .3, .4 and 6.44.8 it follows that

$$\overline{f} \in \text{grandpave } W.$$

Hence, according to 6.44.9,

$$\overline{f} \in \text{greatpave } W.$$

Letting

$$M = \text{mbl } \overline{f},$$

we infer that

$$\overline{f} \in \text{pave } \varphi \ W$$

from .4, 6.3.5, 6.73, 6.78 and 6.50.

From 6.79 and 5.6 we have at once

6.80 THEOREM. If

$$F \in \text{net}'', \quad K = \text{netsum } F,$$
$$\varphi = \text{count } \sigma F, \quad W = \text{netweb } F,$$

then

$$\overline{\text{mr}}\ K\varphi \subset \text{greatpave } W \cap \text{pave } \varphi\ W.$$

6.81 THEOREM. If

$$f \in \text{Msr } \sigma F \cap \text{subadd } F,$$
$$g = \text{strc } f\ F \in \text{netsum } F,$$
$$\varphi = \text{count } \sigma F, \quad \overline{g} = \overline{\text{m}}\ g\ \varphi,$$

then

$$f(\beta) = \overline{g}(\beta) \text{ whenever } \beta \in \text{mbl } \overline{g}.$$

Proof. We let

$$F' = F \cap \text{dmn}' f, \quad S = \sigma F$$

and divide the remainder of the proof into six parts, the first two of which are obvious.

Part 1. $f(\beta) \leqslant \overline{g}(\beta)$ whenever $\beta \subset S$.

Part 2. $f(\beta) = \overline{g}(\beta)$ whenever $\beta \in F$.

Part 3. If $B \in F'$ and $\beta \in \text{mbl } \overline{g}$, then

$$f(B\beta) = \overline{g}(B\beta).$$

Proof. Use Parts 1 and 2 to check that

$$\begin{aligned} 0 &\leqslant \overline{g}(B\beta) - f(B\beta) \\ &\leqslant \overline{g}(B\beta) - f(B\beta) + \overline{g}(B \sim\beta) - f(B \sim\beta) \\ &= \overline{g}(B\beta) + \overline{g}(B \sim\beta) - (f(B\beta) + f(B \sim\beta)) \\ &= \overline{g}(B) - (f(B\beta) + f(B \sim\beta)) \\ &\leqslant \overline{g}(B) - f(B) = 0. \end{aligned}$$

Part 4. If $B \in F'$ and $\beta \in \text{mbl}\,\overline{g}$, then

$$f(B\beta) + f(B \sim\beta) = f(B).$$

Proof. Use Parts 1 and 2 to check that

$$\begin{aligned} f(B) &\leqslant f(B\beta) + f(B \sim\beta) \\ &\leqslant \overline{g}(B\beta) + \overline{g}(B \sim\beta) = \overline{g}(B) = f(B). \end{aligned}$$

Part 5. If $\beta \in \text{mbl}\,\overline{g}$ and $G \in \text{dsjn}''\,F'$, then

$$f(\beta\,\sigma G) \geqslant \Sigma\, B \in G\, f(\beta B).$$

Proof. The desired conclusion is a consequence of the

Statement. If H is a finite subfamily of G, then

$$f(\sigma H\,\beta) = \Sigma\, B \in H\, f(B\beta).$$

Proof. Use Part 4 and 6.2.4 to check that

$$\begin{aligned} 0 &\leqslant \Sigma\, B \in H\, f(B\beta) - f(\sigma H\,\beta) \\ &\leqslant \Sigma\, B \in H\, f(B\beta) - f(\sigma H\,\beta) + \Sigma\, B \in H\, f(B \sim\beta) - f(\sigma H \sim\beta) \\ &= \Sigma\, B \in H\, (f(B\beta) + f(B \sim\beta)) - (f(\sigma H\,\beta) + f(\sigma H \sim\beta)) \\ &= \Sigma\, B \in H\, f(B) - (f(\sigma H\,\beta) + f(\sigma H \sim\beta)) \\ &\leqslant \Sigma\, B \in H\, f(B) - f(\sigma H) \leqslant 0. \end{aligned}$$

Part 6. If $\beta \in \text{mbl}\,\overline{g}$, then

$$f(\beta) = \overline{g}(\beta).$$

Proof. Use 6.67.1 and 6.65.2 so to choose

$$G \in \text{dsjn}''\, F'$$

that

$$\sigma G = S.$$

Using parts 5, 3 and 1, we now see

$$\begin{aligned} f(\beta) &= f(\sigma G \beta) \geqslant \Sigma\, B \in G\, f(B\beta) \\ &= \Sigma\, B \in G\, \overline{g}(B\beta) \geqslant \overline{g}(\beta) \geqslant f(\beta). \end{aligned}$$

With 6.81 in mind, we formulate

6.82 DEFINITION. netmeasure $F = \mathsf{E}\, f \in \text{Msr}\ \sigma F$ (

$$F \in \text{net}'',\ f \in \text{underadd}\ F,\ \text{strc}\ f\ F \in \text{netsum}\ F,$$
$$f \in \text{Hull mbl}\ \overline{\mathsf{m}}\ \text{strc}\ f\ F\ \text{count}\ \sigma F).$$

6.83 THEOREM. If $F \in \text{net}''$,

$$K = \text{netsum}\ F,\quad \varphi = \text{count}\ \sigma F,\quad W = \text{netweb}\ F,$$

then

$$\text{netmeasure}\ F = \overline{\text{mr}}\, K\, \varphi \subset \text{greatpave}\ W \cap \text{pave}\ \varphi\ W.$$

Proof. We let $S = \sigma F$ and divide the remainder of the proof into three parts.

Part 1. If

$$f \in \text{netmeasure } F, \;\; g = \text{strc } f\, F, \;\; \overline{g} = \overline{\text{m}}\, g\, \varphi,$$

then

$$f = \overline{g} \in \overline{\text{mr}}\, K \varphi.$$

Proof. Since $g \in$ netsum F, the desired conclusion is a consequence of 5.6 and the

Statement. If $\beta \subset S$, then

$$f(\beta) = \overline{g}(\beta).$$

Proof. So choose

$$\beta' \in \text{sp}\, \beta \cap \text{mbl}\, \overline{g}$$

that

$$f(\beta) = f(\beta')$$

and use 6.81 and 6.3.2 in checking

$$f(\beta) = f(\beta') = \overline{g}(\beta') \geqslant \overline{g}(\beta) \geqslant f(\beta).$$

Part 2. If $f \in \overline{\text{mr}}\, K \varphi$, then

$$f \in \text{netmeasure } F.$$

Proof. So choose

$$h \in \text{netsum } F$$

that

$$f = \overline{\mathsf{m}}\, h\, \varphi$$

and let

$$g = \text{strc } f\, F, \;\; \overline{g} = \overline{\mathsf{m}}\, g\, \varphi.$$

According to 6.77,

$$F \subset \text{mbl } f$$

and hence, because of 6.3.5, 6.3.9, 6.72 and 6.3.10,

.1 $$f \in \text{underadd } F, \;\; g \in \text{underadd } F, \;\; \text{dmn}'\, h \subset \text{dmn}'\, g$$

and

.2 $$g \in \text{netsum } F.$$

Again according to 6.77,

$$F \subset \text{mbl } \overline{g}.$$

Consequently,

$$\text{Meet}''\ \text{Join}''\ F \subset \text{mbl } \overline{g}$$

and we can now check

.3 $$f \in \text{Hull Meet}''\ \text{Join}''\ F \subset \text{Hull mbl } \overline{g}.$$

With the help of .1, .2, .3 and 6.82 we conclude

$$f \in \text{netmeasure } F.$$

Part 3. netmeasure $F = \overline{\text{mr}}\, K\varphi \subset$ greatpave $W \cap$ pave $\varphi\, W$.

Proof. Use Part 1, Part 2 and Theorem 6.80.

With interval functions in mind we formulate

6.84 DEFINITIONS.

.1 M is complemental if and only if $\sigma M \sim \beta \in$ Join$''$ M whenever $\beta \in M$.

.2 thin K = E $\beta \in$ meet$'$ K (sb $\beta \cap K = 0$).

.3 bloc = E K (underadd $K \subset$ undersum K, K is complemental, meet$'$ K ~thin $K \subset$ Join$''$ K).

.4 mist K = The function φ on sb σK such that, for each $\beta \in$ dmn φ, $\varphi(\beta)$ is 0 or ∞ according to whether β is or is not in sbmb Join$''$ thin K.

.5 massive K = E $\varphi \in$ Msr σK ($K \subset$ mbl φ).

.6 blocmeasure K = E $\varphi \in$ massive K (thin $K \subset$ Zr φ).

.7 blocmass $K\,W$ = E $\varphi \in$ blocmeasure K (Weball W dmn$'$ φ).

.8 subgauge $K\,W$ = E $f \in$ Subadd ($K \subset$ dmn $f \subset$ meet$'$ K and Weball W dmn$'$ f).

.9 undergauge $K\,W$ = E $f \in$ subgauge $K\,W$ (dmn f = meet$'$ K).

.10 inweb K = E $W \in$ web ($K \in$ bloc, reach $W \subset$ meet$'$ K ~thin K, blocmass $K\,W \subset$ coverpave W, and for each $W' \in$ coweb W and each $B \in$ meet$'$ $K \cap$ sp dmn W' and each $\varphi \in$ blocmass $K\,W$

$$\text{Webalm } \varphi\, W' \text{ sb } B).$$

We should like to stress here that we do not assume reach $W \subset K$. The squareweb of 6.97.4 is quite special.

6.85 LEMMA. If

$$f \in \text{undersum } M,\ \ S = \sigma M,\ \ \xi \in \text{Msr } S,$$
$$\overline{f} = \overline{\mathrm{m}}\, f\, \xi,\ \ A \in M,\ \ S \sim A \in \text{Join}''\, M,\ \ T \in \text{dmn } f,$$

then

$$f(T) \geqslant \overline{f}(TSA) + \overline{f}(TS \sim A).$$

Proof. Let $r > 0$. Choose such a countable subfamily C' of M that

$$S \sim A = \sigma C'$$

and let

$$C = \text{sng } A \cup C'.$$

In accordance with 6.2.5 choose such a countable subfamily K of M that

$$K \subset\subset C,\ \ \sigma K = \sigma C,\ \ f(T) + r \geqslant \Sigma\ \beta \in K\, f(T\beta).$$

Note that

$$\sigma K = S.$$

Now if $\beta \in K$, then:

$$\beta \subset A \text{ or } \beta \subset \sim A;$$
$$\overline{f}(T\beta) = \overline{f}(T\beta A) + \overline{f}(T\beta \sim A).$$

Hence

$$\begin{aligned} f(T) + r &\geqslant \Sigma\, \beta \in K\, f(T\beta) \geqslant \Sigma\, \beta \in K\, \overline{f}\,(T\beta) \\ &= \Sigma\, \beta \in K(\overline{f}\,(T\beta A) + \overline{f}\,(T\beta{\sim}A)) \\ &= \Sigma\, \beta \in K\,\overline{f}\,(T\beta A) + \Sigma\, \beta \in K\,\overline{f}\,(T\beta{\sim}A) \\ &\geqslant \overline{f}\,(T\,\sigma K\, A) + \overline{f}\,(T\,\sigma K\,{\sim}A) \\ &= \overline{f}\,(TSA) + \overline{f}\,(TS{\sim}A). \end{aligned}$$

The desired conclusion is at hand.

Because of 2.14 with $C = S$ and because of 6.85, we now have at once

6.86 THEOREM. If

$$f \in \text{undersum } M, \quad S = \sigma M, \quad \xi \in \text{Msr } S,$$
$$\overline{f} = \overline{\mathsf{m}}\, f\, \xi, \quad A \in M, \quad S \sim A \in \text{Join}''\, M,$$

then

$$A \in \text{mbl}\, \overline{f}.$$

6.87 THEOREM. If M is complemental,

$$f \in \text{undersum } M, \quad \xi \in \text{Msr } \sigma M,$$

then

$$\overline{\mathsf{m}}\, f\, \xi \in \text{massive } M.$$

6.88 THEOREM. If K is a family of sets and $\xi = \text{mist } K$, then

$$\xi \in \text{Msr } \sigma K \quad \text{and} \quad \text{thin } K \subset \text{Zr } \xi.$$

6.89 THEOREM. If K is complemental,

$$f \in \text{undersum } K, \quad \xi = \text{mist } K,$$

then

$$\overline{\mathsf{m}}\, f\, \xi \in \text{blocmeasure } K.$$

6.90 THEOREMS.

.1 undergauge $K\ W \subset$ subgauge $K\ W$.

.2 If

$$f \in \text{subgauge } K\ W, \quad \xi \in \text{Msr } \sigma K, \quad \overline{f} = \overline{\mathsf{m}}\, f\, \xi,$$

then

$$\text{dmn}'\, f \subset \text{dmn}'\, \overline{f} \quad \text{and} \quad \text{Weball } W\ \text{dmn}'\, \overline{f}$$

.3 $K \subset \text{meet}'\ K \sim \text{thin } K$.

.4 If $K \in$ bloc, then $K \subset \text{meet}'\ K \sim \text{thin } K \subset \text{Join}''\ K$.

.5 undergauge $K\ W \subset$ underadd meet$'\ K$

$$\subset \text{underadd } (\text{meet}'\ K \sim \text{thin } K)$$
$$\subset \text{underadd } K.$$

Proof. To infer the first inclusion use 6.3.3 and the fact that

$$\text{meet}'\ K \cap \cap\ \text{meet}'\ K = \text{meet}'\ K.$$

To infer the other inclusions use 6.3.9.

.6 If $K \in$ bloc, then

$$\text{undergauge } K\ W \subset \text{undersum } K \subset \text{undersum } (\text{meet}'\ K \sim \text{thin } K).$$

Proof. Use .5 and 6.3.11.

6.91 THEOREM. If

$$K \in \text{bloc} \quad \text{and} \quad f \in \text{undergauge } K\ W,$$

then

$$\overline{\mathsf{m}}\, f \text{ mist } K \in \text{blocmass } K\ W.$$

6.92 THEOREM. If

$$W \in \text{inweb } K \quad \text{and} \quad \varphi \in \text{blocmass } K\ W,$$

then

$$\text{undergauge } K\ W \subset \text{Underpave } \varphi\ W.$$

Proof. Suppose

.1 $$f \in \text{undergauge } K\ W$$

and

$$M = \text{meet}'\ K \sim \text{thin } K.$$

Because of 6.84.10 we may take $C = B$ in 6.44.1 in checking

$$\text{coweb } W \subset \text{Underweb } f\ \varphi.$$

Hence because of 6.46.2

.2 $$f \in \text{Undertile } \varphi\ W.$$

Because of 6.84.10 and 6.9.3

.3 $\varphi \in$ coverpave $W \subset$ covertile W.

Since

$$\text{Weball } W \text{ dmn}' f,$$

it is clear from .3, 3.12, 4.3.1 that

.4 $f \in$ pretile φ W.

Because of 6.90.6 and 6.3.12,

.5 $f \in$ undersum $M \cap$ subadd M.

Because of 6.84.10 and 3.12.9,

$$\text{Weball } W\, M$$

and hence

.6 Webalm φ W M.

From .2, .3, .4, .5, .6 and 6.44.7 we infer

$$f \in \text{Underpave } \varphi\, W,$$

and from this and the arbitrary nature of f in .1 we conclude the desired result.

The properties of subgauge K W stem in large measure from 6.3.14 and

6.93 THEOREM. If

$$K \in \text{bloc}, \;\; g \in \text{subgauge } K\, W,$$
$$g \subset f \in \text{undergauge } K\, W,$$
$$\xi = \text{mist } K, \;\; \overline{g} = \overline{\mathrm{m}}\, g\, \xi, \;\; \overline{f} = \overline{\mathrm{m}}\, f\, \xi,$$

then

$$\overline{g} = \overline{f} \in \text{blocmass } K\, W.$$

We divide the proof into four parts.

Part 1. If $B \in \text{meet}'\, K$, then $\overline{g}(B) \leqslant f(B)$.

Proof. If $B \in \text{thin } K$, then $\overline{g}(B) = 0$. Accordingly we assume

$$B \in \text{meet}'\, K \sim \text{thin } K.$$

Let $r > 0$ and choose such a countable subfamily C of K that

$$B = \sigma C$$

and in accordance with 6.2.5 choose such a countable subfamily H of K that

$$H \subset\subset C, \;\; \sigma H = \sigma C, \;\; f(B) + r \geqslant \Sigma\, \beta \in H\, f(B\beta).$$

We now have

$$\begin{aligned} f(B) + r &\geqslant \Sigma\, \beta \in H\, f(B\beta) = \Sigma\, \beta \in H\, f(\beta) = \Sigma\, \beta \in H\, g(\beta) \\ &\geqslant \Sigma\, \beta \in H\, \overline{g}(\beta) \geqslant \overline{g}(\sigma H) = \overline{g}(\sigma C) = \overline{g}(B). \end{aligned}$$

Part 2. If $B \in \text{dmn } \xi$, then $\overline{g}(B) \leqslant \overline{f}(B)$.

Proof. Use Part 1 and 2.12.5.

Because of 2.12.7 we have

Part 3. If $B \in \operatorname{dmn} \xi$, then $\overline{f}(B) \leqslant \overline{g}(B)$.

Part 4. $\overline{g} = \overline{f} \in$ blocmass K W.

Proof. From Parts 2 and 3 we learn

$$\overline{g} = \overline{f}$$

and from 6.91 we learn

$$\overline{f} \in \text{blocmass } K \; W.$$

Stronger than 6.91 is

6.94 THEOREM. If

$$K \in \text{bloc} \quad \text{and} \quad g \in \text{subgauge } K \; W,$$

then

$$\overline{\mathrm{m}} \; g \text{ mist } K \in \text{blocmass } K \; W.$$

Proof. In accordance with 6.3.14 so choose f that

$$g \subset f \in \text{undergauge } K \; W,$$

let

$$\xi = \text{mist } K, \; \overline{g} = \overline{\mathrm{m}} \; g \; \xi, \; \overline{f} = \overline{\mathrm{m}} \; f \; \xi,$$

and use 6.93 to conclude

$$\overline{\mathrm{m}} \; g \text{ mist } K = \overline{\mathrm{m}} \; g \; \xi = \overline{g} = \overline{f} \in \text{blocmass } K \; W.$$

Akin to 6.92 is

6.95 THEOREM. If

$$W \in \text{inweb } K \text{ and } \varphi \in \text{blocmass } K\, W,$$

then

$$\text{subgauge } K\, W \subset \text{tile } \varphi\, W.$$

Proof. Suppose

$$g \in \text{subgauge } K\, W$$

and again in accordance with 6.3.14 choose f so that

$$g \subset f \in \text{undergauge } K\, W.$$

According to 6.92, 6.51 and 6.20.1,

$$\begin{aligned}\text{undergauge } K\, W &\subset \text{Underpave } \varphi\, W \\ &\subset \text{subpave } \varphi\, W \subset \text{tile } \varphi\, W.\end{aligned}$$

Hence

$$\begin{aligned}&f \in \text{tile } \varphi\, W, \\ &\text{dmn}'\, g \subset \text{dmn } g \subset \text{dmn } f \subset \text{tiled } f\, \varphi\, W,\end{aligned}$$

and 4.14 assures us

$$g \in \text{tile } \varphi\, W.$$

6.96 THEOREM. If

$$W \in \text{inweb } K, \;\; g \in \text{subgauge } K\, W, \;\; h \in \text{subgauge } K\, W,$$
$$\overline{h} = \overline{\mathsf{m}}\, h \text{ mist } K,$$

then

$$\text{Al } \overline{h} \text{ dmn } W\, x\; (0 \leqslant \mathsf{D}\, W\, g\, h\, x < \infty).$$

Proof. Let $\varphi = \overline{h}$ and use 2.12.6 in checking

.1 $$\overline{h} = \overline{\mathsf{m}}\, h\, \overline{h} = \overline{\mathsf{m}}\, h\, \varphi.$$

According to 6.94

$$\varphi \in \text{blocmass } K\, W$$

and hence according to 6.95

.2 $$g \in \text{tile } \varphi\, W \;\text{ and }\; h \in \text{tile } \varphi\, W.$$

The desired conclusion now follows from .1, .2 and 5.3.

6.97 DEFINITIONS.

.1 rf $= \mathsf{E}\, x\; (-\infty < x < \infty)$.

.2 pl $= \mathsf{E}\, x,y\; (x \in \text{rf and } y \in \text{rf})$.

.3 intervalweb = The function W on rf for which $W(x) = \mathsf{E}\, r,\beta\; [0 < r < \infty$ and β is of the form

$$\mathsf{E}\, t\; (a \leqslant t \leqslant b),$$

where $-\infty < a \leqslant x \leqslant b < \infty$ and

$$0 < b - a \leqslant r]$$

whenever $x \in \mathrm{rf}$.

.4 squareweb = The function W on pl for which $W(x,y) = \mathsf{E}\, r,\beta\ [0 < r < \infty$ and β is of the form

$$\mathsf{E}\, s,t\ (x - h \leqslant s \leqslant x + h \ \text{ and } \ y - h \leqslant t \leqslant y + h),$$

where

$$0 < h \leqslant r]$$

whenever $x \in \mathrm{rf}$ and $y \in \mathrm{rf}$.

As we have said before, in connection with 6.84.10, squareweb is quite special. Our theory applies to many other rectangular webs, but in such applications the K of 6.84 would usually be the reach of squareweb.

Theorems 6.98.1 and 6.98.2 interest us. We found the first easier to prove directly than the second. They both follow from the work of D. C. Peterson[3], who has shown among other things that if $1 \leqslant n \in \omega$ and K is the family of ordinary n-cubes of positive finite diameter, then

$$\text{underadd } K \subset \text{undersum } K.$$

In connection with 6.98.3, let

$$\Delta = \varphi = \text{sct } \psi\ B$$

and check that 6.4 and 6.12 of *Covering and Differentiation*[10] may be used to see that

$$B \in \text{coverpaved } \psi\ W.$$

In connection with 6.98.4, use 5.13 of *Perfect Blankets*[5] in checking that

$$B \in \text{coverpaved } \psi \ W.$$

In connection with 6.98.5, use problem 1.J (d) of Kelley[11] to see that

$$\text{blocmass } K \ W \subset \text{covertile } W$$

and then use 6.98.3 in checking that

$$\text{blocmass } K \ W \subset \text{coverpave } W.$$

In connection with 6.98.6, use the Lindelöf Theorem to see that

$$\text{blocmass } K \ W \subset \text{covertile } W$$

and then use 6.98.4 in checking

$$\text{blocmass } K \ W \subset \text{coverpave } W.$$

6.98 THEOREMS.

.1 If K = reach intervalweb, then:

underadd $K \subset$ undersum K;

$K \in$ bloc;

Zr mist K = the family of countable subsets of rf.

.2 If K = reach squareweb, then:

underadd $K \subset$ undersum K;

$K \in$ bloc;

Zr mist K = the family of those $\beta \subset$ pl such that for some T

$$\text{dmn } (\beta T) \cup \text{rng } (\beta \sim T) \text{ is countable.}$$

.3 If

$$W \in \text{coweb intervalweb}, \quad K = \text{reach intervalweb},$$
$$\psi \in \text{blocmass } K \ W, \quad B \in \text{dmn}' \ \psi,$$

then

$$B \in \text{coverpaved } \psi\ W.$$

.4 If

$$W \in \text{coweb squareweb}, \quad K = \text{reach squareweb},$$
$$\psi \in \text{blocmass } K\ W, \quad B \in \text{dmn}'\ \psi,$$

then

$$B \in \text{coverpaved } \psi\ W.$$

.5 If

$$W \in \text{coweb intervalweb}, \quad K = \text{reach intervalweb},$$

then

$$W \in \text{inweb } K.$$

.6 If

$$W \in \text{coweb squareweb}, \quad K = \text{reach squareweb},$$

then

$$W \in \text{inweb } K.$$

6.99 DEFINITION. $\mathsf{D}\, f/g\ x = \lim_{t\to 0} \dfrac{f(x+t) - f(x)}{g(x+t) - g(x)}.$

In particular we have

6.100 THEOREM. If K is the family of ordinary closed intervals of finite positive length, g and h are complex valued functions of bounded variation on each member of K, w is the variation function of h, V is such a function on K that

$$V(\beta) = w(b) - w(a)$$

whenever $-\infty < a < b < \infty$ and

$$\beta = \mathsf{E}\, t\ (a \leqslant t \leqslant b),$$

$\overline{h} = \overline{\mathsf{m}}\ V$ mist K, then

.1 Alm $\overline{h}\ x\ (\,|\mathsf{D}\ g\,/\,w x| < \infty)$,

.2 Alm $\overline{h}\ x\ (\,|\mathsf{D}\ h\,/\,w x\,| = 1\,)$,

.3 Alm $\overline{h}\ x\ (\,|\mathsf{D}\ g\,/\,h\ x\ | < \infty)$.

Because of 6.98.5 and 6.96 it is easy to see .1. With some additional effort the precise estimate in .2 can be verified directly. Now .3 is obvious by division.

In connection with 6.100, it is easy to check that if C is the set of points of continuity of h and if

$$\overline{w} = \overline{\mathsf{m}}\ V \text{ count rf},$$

then

$$\overline{h} = \text{sct}\ \overline{w}\ C.$$

To us 6.100 is of considerable interest when h is a Cantor-like singular function.

7. THE INTEGRATION OF THE DERIVATIVE AND THE DIFFERENTIATION OF THE INTEGRAL.

From 6.13.7, 6.20.3, 4.5.4, 4.5.3, 5.5 and 6.11.4 we infer

7.1 THEOREM. If

$$\varphi \in \text{solidtile } W, \ f \in \text{tile } \varphi \ W,$$

then

$$\text{Alm } \varphi \ x \ (0 \leqslant \mathsf{D} \ W \ f \ \varphi \ x < \infty).$$

7.2 THEOREM. If

$$\varphi \in \text{solidtile } W, \ f \in \text{tile } \varphi \ W, \ \overline{f} = \overline{\mathsf{m}} \ f \ \varphi,$$

then

$$\overline{f} \in \text{Hull mbl } \varphi.$$

Proof. Let

$$S = \text{rlm } \varphi, \ F = \text{mbl } \varphi.$$

The desired conclusion is a consequence of the

Statement. If

$$C \subset S, \ r > 0,$$

then there is a

$$C' \in F \cap \operatorname{sp} C$$

for which

$$\overline{f}(C') \leqslant \overline{f}(C) + r.$$

Proof. Use 6.20.10 and 4.17 to secure such a

$$G \in \operatorname{cvr} \varphi\, C\, F$$

that

$$\Sigma\, \beta \in G\, f(\beta) \leqslant \overline{f}(C) + r.$$

Let

$$C' = \sigma G \cup C \sim \sigma G.$$

Clearly

$$C' \in F \cap \operatorname{sp} C.$$

Also since

$$C \sim \sigma G \in \operatorname{Zr} \varphi$$

we know

$$\overline{f}(C \sim \sigma G) = 0.$$

We now conclude

$$\begin{aligned} \overline{f}(C) + r &\geqslant \Sigma\, \beta \in G\, f(\beta) \geqslant \Sigma\, \beta \in G\, \overline{f}(\beta) \\ &\geqslant \overline{f}(\sigma G) = \overline{f}(\sigma G) + \overline{f}(C \sim \sigma G) \geqslant \overline{f}(C'). \end{aligned}$$

7.3 LEMMA. If

$$\varphi \in \text{solidtile } W, \ f \in \text{tile } \varphi \ W,$$
$$\beta \in \text{mbl}' \ \varphi, \ -\infty < \lambda < \infty,$$

then

$$\beta \ \mathsf{E} \ x \ (\mathsf{D} \ W f \varphi x > \lambda) \in \text{mbl } \varphi.$$

Proof. If $-\infty < \lambda < 0$, then the desired conclusion is an immediate consequence of 7.1. We assume therefore that $0 \leqslant \lambda < \infty$ and let

$$\overline{f} = \overline{\mathsf{m}} \ f \varphi, \quad T = \beta \ \mathsf{E} \ x \ (\mathsf{D} \ W f \varphi x > \lambda), \quad B = \beta \sim T.$$

Also let ξ and A be such sequences that

$$\xi(n) = \lambda + 1/(n+1), \ \ A(n) = \beta \ \mathsf{E} \ x \ (\mathsf{D} \ W f \varphi x > \xi(n))$$

whenever $n \in \omega$. Next with the help of 6.20.7 we so secure

$$B' \in \text{sp } B \cap \text{mbl } \varphi$$

that

$$\varphi(B') = \varphi(B)$$

and divide the remainder of the proof into three parts.

Part 1. Al $\varphi \ B \ x \ (\mathsf{D} \ W f \varphi x \leqslant \lambda)$.

Proof. Use 7.1.

Part 2. If $n \in \omega$, then $\varphi(A(n) \cap B') = 0$.

Proof. Let $S = A(n)$ and so secure

$$S' \in \operatorname{sp} S \cap \operatorname{mbl} \varphi$$

that

$$\varphi(S') = \varphi(S).$$

Since

$$\varphi(\beta) < \infty$$

we infer from 2.7, 7.2 and 2.8 that

$$\begin{aligned} \varphi(S'B) &= \varphi(S'B') = \varphi(SB'), \\ \overline{f}(S'B) &= \overline{f}(S'B') = \overline{f}(SB'). \end{aligned}$$

It now follows from 6.11.4, 6.20.3, 4.3.4, 4.28, 2.13 and Part 1 that

$$\begin{aligned} \xi(n) \cdot \varphi(SB') \leqslant \overline{f}(SB') = \overline{f}(S'B) &\leqslant \lambda \cdot \varphi(S'B) \\ &= \lambda \cdot \varphi(SB') \leqslant \lambda \cdot \varphi(\beta) < \infty. \end{aligned}$$

Hence

$$(\xi(n) - \lambda) \cdot \varphi(SB') \leqslant 0.$$

Accordingly, since

$$\xi(n) - \lambda > 0,$$

it now follows that

$$0 \leqslant \varphi(SB') \leqslant 0.$$

The desired conclusion is at hand.

Part 3. $T \in \text{mbl}\ \varphi$.

Proof. Since Part 2 assures us

$$\varphi(TB') \leqslant \Sigma\ n \in \omega\ \varphi(A(n) \cap B') = 0$$

we now know

$$TB' \in \text{mbl}\ \varphi.$$

Moreover

$$\begin{aligned} &\beta \sim T = B \subset B',\ \ \beta \sim T \sim B' = 0, \\ &T = TB' \cup T \sim B' = TB' \cup \beta T \sim B' \cup 0 \\ &\ \ = TB' \cup \beta T \sim B' \cup \beta \sim T \sim B' \\ &\ \ = TB' \cup \beta \sim B' \in \text{mbl}\ \varphi. \end{aligned}$$

7.4 THEOREM. If

$$\varphi \in \text{solidtile}\ W,\ \ f \in \text{tile}\ \varphi\ W,$$

then

$$\mathsf{D}\ W f\ \varphi\ x \text{ is } \varphi \text{ measurable in } x.$$

Proof. The desired conclusion is a consequence of 7.1 and the

Statement. If $-\infty < \lambda < \infty$,

then

$$\mathsf{E}\ x\ (\mathsf{D}\ W f\ \varphi\ x > \lambda) \in \text{mbl}\ \varphi.$$

Proof. Let

$$C = \mathsf{E}\, x\, (\mathsf{D}\ W f\, \varphi\, x > \lambda), \quad F = \text{mbl}'\, \varphi, \quad \psi = \varphi.$$

From 7.3, 6.20.10 and 4.29.2 we learn

$$C \in \text{mbl}\ \varphi.$$

7.5 LEMMA. If

$$\varphi \in \text{solidtile}\ W, \quad f \in \text{tile}\ \varphi\ W,$$
$$\overline{f} = \overline{\mathsf{m}}\, f\, \varphi, \quad 0 < p < q < \infty,$$
$$C = \mathsf{E}\, x\, (p < \mathsf{D}\ W f\, \varphi\, x < q),$$

then

$$\text{mbl}\ \varphi \subset \text{mbl sct}\ \overline{f}\, C.$$

Proof. Let

$$\psi = \text{sct}\ \overline{f}\ C$$

and notice that the desired conclusion is a consequence of the

Statement. If

$$A \in \text{mbl}\ \varphi, \quad T \in \text{dmn}'\ \psi, \quad 0 < r < \infty,$$

then

$$\psi(T) + r \geqslant \psi(TA) + \psi(T \sim A).$$

Proof. Let

$$3 \cdot r' = r, \ M = \psi(T) + r'$$

and

$$\xi = p \cdot r'/(q \cdot M + 3 \cdot p \cdot r'),$$

so that

.1 $$0 < 2 \cdot \xi < 1$$

and

.2 $$q \cdot \xi \cdot M/p \leqslant r'$$

Let

.3 $$H_1 = \mathsf{E}\, \beta(\varphi(\beta \sim A) < \xi \cdot \varphi(\beta)),$$

.4 $$H_2 = \mathsf{E}\, \beta(\varphi(\beta A) < \xi \cdot \varphi(\beta)),$$

.5 $$H = H_1 \cup H_2,$$

.6 $$K = \mathsf{E}\, \beta(p \cdot \varphi(\beta) < f(\beta)),$$

.7 $$F = H \cap K,$$

.8 $$W' = \text{strc } W\, C,$$

$$W_1 = \text{strc } W\, A, \ W_2 = \text{strc } W \sim A.$$

From 5.17 and 3.14.2 we learn

$$\text{knap } W_1 \sim H_1 \in \text{Zr } \varphi, \ \text{knap } W_2 \sim H_2 \in \text{Zr } \varphi.$$

Consequently

$$\text{knap } W_1 \sim H \in \text{Zr } \varphi, \ \text{knap } W_2 \sim H \in \text{Zr } \varphi.$$

Thus because of 3.13.2

$$\text{knap } W \sim H \in \text{Zr } \varphi,$$

and consequently

$$\text{Webalm } \varphi \; W \; H.$$

Accordingly, because of .8,

.9 $$\text{Webalm } \varphi \; W' \; H.$$

Now, because of 3.14.4,

.10 $$\text{knap } W' \sim K = 0, \;\; \text{Weball } W' \; K,$$
$$\text{Webalm } \varphi \; W' \; K.$$

From .9, .10 and .7 we infer

$$\text{Webalm } \varphi \; W' \; F$$

and since

$$T \, C \subset C = \text{dmn } W',$$

we can and do use 4.17 and 4.6.5 to secure such a

$$G \in \text{cvr } \varphi(T \, C) \; F$$

that

.11 $$\Sigma \; \beta \in G \; f(\beta) \leqslant \overline{f}(T \, C) + r'.$$

We now let

.12 $$G_1 = G \cap H_1, \quad G_2 = G \cap H_2,$$

and divide the remainder of the proof into seven parts.

Part 1. $\Sigma\ \beta \in G\ \varphi(\beta) \leqslant M/p$.

Proof. Helped by .7 and .6 we infer

$$M \geqslant \Sigma\ \beta \in G\ f(\beta) \geqslant \Sigma\ \beta \in G\,(p \cdot \varphi(\beta)) = p \cdot \Sigma\ \beta \in G\ \varphi(\beta).$$

Part 2. $G = G_1 \cup G_2$ and $G_1 \cap G_2 = 0$.

Proof. The first conclusion follows from .12, .5, .7 and our choice of G. The second conclusion follows from the fact that .3, .4 and .1 insure

$$H_1 \cap H_2 = 0.$$

Part 3. $\varphi(TCA \sim \sigma G_1) \leqslant \xi \cdot M/p$.

Proof. Since

$$\begin{aligned} TCA \sim \sigma G_1 \sim \sigma G_2 \subset TC \sim \sigma G_1 \sim \sigma G_2 &= TC \sim (\sigma G_1 \cup \sigma G_2) \\ &= TC \sim \sigma G \in \mathrm{Zr}\ \varphi, \end{aligned}$$

we are now sure

$$TCA \sim \sigma G_1 \in \mathrm{sb}'\ \varphi\ \sigma G_2.$$

Consequently, with the help of .12, .4 and Part 1 we infer

$$\begin{aligned} \varphi(TCA \sim \sigma G_1) &\leqslant \varphi(TCA\ \sigma G_2) \leqslant \Sigma\ \beta \in G_2\ \varphi(TCA\beta) \\ &\leqslant \Sigma\ \beta \in G_2\ \varphi(\beta A) \leqslant \Sigma\ \beta \in G_2\ (\xi \cdot \varphi(\beta)) \\ &= \xi \cdot \Sigma\ \beta \in G_2\ \varphi(\beta) \leqslant \xi \cdot \Sigma\ \beta \in G\ \varphi(\beta) \\ &\leqslant \xi \cdot M/p. \end{aligned}$$

Part 4. $\varphi(TC\sim A\sim \sigma G_2)\leqslant \xi\cdot M/p$.

Proof. Since

$$TC\sim A\sim \sigma G_1\sim \sigma G_2\subset TC\sim \sigma G_1\sim \sigma G_2 = TC\sim(\sigma G_1\cup \sigma G_2)$$
$$= TC\sim \sigma G\in \mathrm{Zr}\ \varphi,$$

we are now sure

$$TC\sim A\sim \sigma G_2\in \mathrm{sb}'\ \varphi\ \sigma G_1.$$

Consequently, with the help of .12, .3 and Part 1 we infer

$$\begin{aligned}\varphi(TC\sim A\sim \sigma G_2) &\leqslant \varphi(TC\sim A\ \sigma G_1)\leqslant \Sigma\ \beta\in G_1\ \varphi(TC\sim A\ \beta)\\ &\leqslant \Sigma\ \beta\in G_1\ \varphi(\beta\sim A)\leqslant \Sigma\ \beta\in G_1\ (\xi\cdot\varphi(\beta))\\ &= \xi\cdot \Sigma\ \beta\in G_1\ \varphi(\beta)\leqslant \xi\cdot\Sigma\ \beta\in G\ \varphi(\beta)\\ &\leqslant \xi\cdot M/p.\end{aligned}$$

Part 5. $\overline{f}(TCA\sim \sigma G_1)\leqslant r'$.

Proof. From 4.28.1, Part 3 and .2 we infer

$$\overline{f}(TCA\sim\sigma G_1)\leqslant q\cdot\varphi(TCA\sim\sigma G_1)\leqslant q\cdot\xi\cdot M/p\leqslant r'.$$

Part 6. $\overline{f}(TC\sim A\sim \sigma G_2)\leqslant r'$.

Proof. From 4.28.1, Part 4 and .2 we infer

$$\overline{f}(TC\sim A\sim\sigma G_2)\leqslant q\cdot\varphi(TC\sim A\sim\sigma G_2)\leqslant q\cdot\xi\cdot M/p\leqslant r'.$$

Part 7. $\psi(T)+r\geqslant \psi(TA)+\psi(T\sim A)$.

Proof. We use .11, Part 2, Part 5 and Part 6 in checking

$$\begin{aligned}\psi(T) + r = \overline{f}(TC) + r &= \overline{f}(TC) + r' + 2 \cdot r' \\ &\geqslant 2 \cdot r' + \Sigma\, \beta \in G\, f(\beta) \\ &= r' + \Sigma\, \beta \in G_1\, f(\beta) + r' + \Sigma\, \beta \in G_2\, f(\beta) \\ &\geqslant r' + \Sigma\, \beta \in G_1\, \overline{f}(\beta) + r' + \Sigma\, \beta \in G_2\, \overline{f}(\beta) \\ &\geqslant r' + \overline{f}(\sigma G_1) + r' + \overline{f}(\sigma G_2) \\ &\geqslant r' + \overline{f}(TCA\ \sigma G_1) + r' + \overline{f}(TC \sim A\ \ \sigma G_2) \\ &\geqslant \overline{f}(TCA \sim \sigma G_1) + \overline{f}(TCA\ \sigma G_1) \\ &\quad + \overline{f}(TC \sim A \sim \sigma G_2) + \overline{f}(TC \sim A\ \ \sigma G_2) \\ &\geqslant \overline{f}(TCA) + \overline{f}(TC \sim A) \\ &= \psi(TA) + \psi(T \sim A).\end{aligned}$$

7.6 THEOREM. If

$$\varphi \in \text{solidtile } W, \ f \in \text{tile } \varphi\ W, \ \overline{f} = \overline{\mathsf{m}}\, f\, \varphi,$$

then

$$\text{mbl } \varphi \subset \text{mbl } \overline{f}.$$

Proof. Let

$$S = \text{rlm } \varphi, \ \ F = \text{mbl } \varphi,$$

and let K be such a sequence that

$$K(n) = \mathsf{E}\, x\ [1/(n+2) < \mathsf{D}\ W\, f\, \varphi\, x < n + 2]$$

whenever $n \in \omega$. Also let

$$Z = S \sim \mathsf{U}\ n \in \omega\ K(n).$$

Because of 7.1 it is clear that

$$\text{Al } \varphi\ Z\ x\ (\mathsf{D}\ W\, f\, \varphi\, x = 0)$$

and hence according to 4.30

$$\overline{f}(Z) = 0.$$

Because of 7.4 and 7.5 we can now replace 'φ' by '$\overline{f}$' in 2.16 and conclude

$$F \subset \text{mbl}\,\overline{f}.$$

With the help of 7.6, 7.2, 4.20 and 6.20.5 we easily check the

7.7 THEOREM. If

$$\varphi \in \text{solidtile } W, \ f \in \text{tile } \varphi \ W, \ \overline{f} = \overline{\mathsf{m}}\, f\, \varphi,$$

then

$$\overline{f} \in \text{measuretile } \varphi \ W.$$

7.8 DEFINITION. $\int A\,;u(x)\,\varphi\,\mathrm{d}x = \int(\mathrm{Cr}\ x\ A \cdot u(x))\,\varphi\,\mathrm{d}x.$

7.9 THEOREM. If

$$\begin{gathered}\varphi \in \text{solidtile } W, \ f \in \text{tile } \varphi \ W, \\ \overline{f} = \overline{\mathsf{m}}\, f\, \varphi, \ B \in \text{mbl } \varphi,\end{gathered}$$

then

$$\overline{f}(B) = \int B\,;(\mathsf{D}\ W\, f\, \varphi\, x)\,\varphi\,\mathrm{d}x.$$

Proof. After recalling 6.20.11 and 6.28.1, let g be the function on mbl φ for which

$$g(\beta) = \int \beta\,;(\mathsf{D}\ W\, f\, \varphi\, x)\,\varphi\,\mathrm{d}x$$

whenever $\beta \in \text{mbl } \varphi$. The desired conclusion is a consequence of the

Statement. If $1 < \lambda < \infty$, then

$$g(B) \leqslant \lambda \cdot \overline{f}(B) \leqslant \lambda \cdot \lambda \cdot g(B).$$

Proof. Let

$$A = B \mathsf{E}\, x\, (0 < \mathsf{D}\, W f \varphi x < \infty),$$
$$C = B \sim A.$$

We recall 2.1.10 and let S be such a function on ω' that

$$S(n) = A \mathsf{E}\, x\, (\lambda^n \leqslant \mathsf{D}\, W f \varphi x < \lambda^{n+1})$$

whenever $n \in \omega'$. Next let

$$H = \mathsf{U}\, n \in \omega' \operatorname{sng}(S(n)),$$
$$G = \operatorname{sng} C \cup H.$$

Evidently $\sigma H = A$ and hence

$$\sigma G = B.$$

Moreover, because of 7.4,

$$G \in \operatorname{dsjn}'' \operatorname{sb} \operatorname{mbl} \varphi.$$

Using 4.28 we see that if

$$n \in \omega' \text{ and } \beta = S(n),$$

then:

$$\lambda^n \cdot \varphi(\beta) \leqslant \overline{f}(\beta) \leqslant \lambda^{n+1} \cdot \varphi(\beta);$$
$$\lambda^n \cdot \varphi(\beta) \leqslant g(\beta) \leqslant \lambda^{n+1} \cdot \varphi(\beta).$$

Since 7.1 assures us

$$\text{Al}\ \varphi\ C\ x\ (\mathsf{D}\ W\ f\ \varphi\ x = 0),$$

and since 4.30 now assures us

$$\overline{f}\,(C) = 0 = g(C),$$

we find it easy to check that

$$g(\beta) \leqslant \lambda \cdot \overline{f}\,(\beta) \leqslant \lambda \cdot \lambda \cdot g(\beta)$$

whenever $\beta \in G$. Consequently, with the help of 7.6 we conclude

$$\begin{aligned} g(B) &= \Sigma\ \beta \in G\ g(\beta) \leqslant \Sigma\ \beta \in G\ (\lambda \cdot \overline{f}\,(\beta)) \\ &= \lambda \cdot \Sigma\ \beta \in G\,\overline{f}\,(\beta) = \lambda \cdot \overline{f}\,(B) \\ &= \Sigma\ \beta \in G\ (\lambda \cdot \overline{f}\,(\beta)) \leqslant \Sigma\ \beta \in G\ (\lambda \cdot \lambda \cdot g(\beta)) \\ &= \lambda \cdot \lambda \cdot \Sigma\ \beta \in G\ g(\beta) = \lambda \cdot \lambda \cdot g(B). \end{aligned}$$

From 7.9 we have at once

7.10 THEOREM. If

$$\varphi \in \text{solidtile}\ W,\ \ f \in \text{tile}\ \varphi\ W,$$

then

$$f(\beta) \geqslant \textstyle\int \beta;\ (\mathsf{D}\ W\ f\ \varphi\ x)\ \varphi\ \mathrm{d}x$$

whenever

$$\beta \in \text{dmn}\ f \cap \text{mbl}\ \varphi.$$

From 7.9 and 4.21 we infer

7.11 THEOREM. If

$$\varphi \in \text{solidtile } W, \ f \in \text{tile } \varphi \ W, \ A \in \text{mbl } \varphi,$$

then a necessary and sufficient condition that

$$A \in \text{contiled } f \ \varphi \ W$$

is that

$$f(A) = \int A ; (\mathsf{D} \ W f \ \varphi \ x) \ \varphi \ \mathrm{d}x.$$

In view of Section 6 and especially 6.37–6.40, our next theorem has many applications.

7.12 THEOREM. If

$$\varphi \in \text{solidtile } W, \ u \in \text{Massable+ } \varphi,$$
$$g = \text{Gr } u \ \varphi \in \text{tile } \varphi \ W,$$

then

$$\text{Alm } \varphi \ x \ (\mathsf{D} \ W \ g \ \varphi \ x = u(x)).$$

Proof. Let

$$F = \text{dmn}' \ g, \ \ \psi = \varphi,$$

and note that

$$F \subset \text{mbl } \varphi, \ \text{Webalm } \varphi \ W \ F.$$

Also let

$$C = \mathsf{E} \ x \in \text{rlm } \varphi \ (\mathsf{D} \ W \ g \ \varphi \ x \neq u(x)).$$

The desired conclusion is a consequence of 4.29.1 and the

Statement. If $\beta \in F$, then

$$\varphi(\beta C) = 0.$$

Proof. We let

$$\overline{g} = \overline{\mathsf{m}}\, g\, \varphi$$

and learn from 6.41.1 that

.1 $$g \subset \overline{g}.$$

Next let

$$\beta' = \beta\, \mathsf{E}\, x\, (\mathsf{D}\, W\, g\, \varphi\, x - u(x) \geqslant 0),\ \ \beta'' = \beta \sim \beta'.$$

Clearly, because of 7.1,

$$\mathrm{Al}\ \varphi\, \beta''\, x\, (u(x) - \mathsf{D}\, W\, g\, \varphi\, x > 0).$$

Because of 7.4, .1 and 7.9,

$$\begin{aligned} 0 &\leqslant g(\beta') = \textstyle\int \beta';\ u(x)\, \varphi\, \mathrm{d}x = \int \beta';\ (\mathsf{D}\, W\, g\, \varphi\, x)\, \varphi\, \mathrm{d}x < \infty, \\ 0 &\leqslant g(\beta'') = \textstyle\int \beta'';\ u(x)\, \varphi\, \mathrm{d}x = \int \beta'';\ (\mathsf{D}\, W\, g\, \varphi\, x)\, \varphi\, \mathrm{d}x < \infty. \end{aligned}$$

Hence

$$\begin{aligned} 0 &= \textstyle\int \beta';\ (\mathsf{D}\, W\, g\, \varphi\, x - u(x))\, \varphi\, \mathrm{d}x + \int \beta'';\ (u(x) - \mathsf{D}\, W\, g\, \varphi\, x)\, \varphi\, \mathrm{d}x \\ &= \textstyle\int \beta';\ |\mathsf{D}\, W\, g\, \varphi\, x - u(x)|\, \varphi\, \mathrm{d}x + \int \beta'';\ |\mathsf{D}\, W\, g\, \varphi\, x - u(x)|\, \varphi\, \mathrm{d}x \\ &= \textstyle\int \beta;\ |\mathsf{D}\, W\, g\, \varphi\, x - u(x)|\, \varphi\, \mathrm{d}x. \end{aligned}$$

Accordingly,

$$\text{Al } \varphi \, \beta \, x \, (\text{D } W \, g \, \varphi \, x = u(x))$$

and

$$\varphi(\beta C) = 0.$$

7.13 THEOREM. If

$$\varphi \in \text{goodtile } W, \quad A = \text{dmn } W, \quad \varphi_1 = \text{sct } \varphi \, A,$$

then

$$\varphi_1 \in \text{solidtile } W.$$

Proof. Let

$$f = \varphi.$$

With the help of 6.20.7 we infer that

$$f \in \text{tile } \varphi \, W$$

and hence, according to 2.13 and 5.8,

$$\varphi_1 = \overline{\text{m}} \, f \, \varphi_1 \in \text{contile } \varphi_1 \, W.$$

Because of 4.5.3 and 4.3.4,

$$\varphi_1 \in \text{fairtile } W.$$

Because of 2.5.5, 2.10.1 and 6.20.7 it is now easy to check that

$$\varphi_1 \in \text{goodtile } W.$$

Since obviously

$$\text{rlm } \varphi_1 = \text{rlm } \varphi \in \text{sb}' \, \varphi_1 \, A \,,$$

we conclude

$$\varphi_1 \in \text{solidtile } W.$$

7.14 THEOREM. If

$$\varphi \in \text{goodtile } W, \; f \in \text{tile } \varphi \, W, \; A = \text{dmn } W,$$

then a necessary and sufficient condition that

$$\mathsf{D} \, W f \varphi \, x \text{ is } \varphi \text{ measurable in } x$$

is that

$$A \in \text{mbl } \varphi.$$

Proof. The necessity of the condition is obvious.

We assume now

$$A \in \text{mbl } \varphi$$

and let

$$\varphi_1 = \text{sct } \varphi \, A \,.$$

According to 7.13 and 4.7.9,

$$\varphi_1 \in \text{solidtile } W, \; f \in \text{tile } \varphi_1 \, W.$$

From 7.4 we now learn

$$\mathsf{D}\ W f \varphi_1\ x \text{ is } \varphi_1 \text{ measurable in } x.$$

But according to 5.14.2,

$$\text{Alm } \varphi_1\ x\ (\mathsf{D}\ W f \varphi_1\ x = \mathsf{D}\ W f \varphi\ x).$$

Hence

$$\mathsf{D}\ W f \varphi\ x \text{ is } \varphi_1 \text{ measurable in } x.$$

Helped by this and 2.5.6 we conclude

$$\mathsf{D}\ W f \varphi\ x \text{ is } \varphi \text{ measurable in } x.$$

7.15 THEOREM. If

$$\varphi \in \text{goodtile } W,\ f \in \text{tile } \varphi\ W,\ A = \text{dmn } W,$$
$$\varphi_1 = \text{sct } \varphi\ A,\ \overline{f} = \overline{\mathsf{m}}\ f \varphi,\ B \in \text{mbl } \varphi_1,$$

then

$$\overline{f}(BA) = \int B; (\mathsf{D}\ W f \varphi\ x)\ \varphi_1\ \mathrm{d}x.$$

Proof. Let

$$F = \text{sct } \overline{f}\ A,\ \overline{f}_1 = \overline{\mathsf{m}}\ f \varphi_1.$$

Since

$$\varphi_1 \in \text{solidtile } W,\ f \in \text{tile } \varphi_1\ W,$$

we conclude from 2.15, 7.9 and 5.14.2 that

$$\overline{f}(BA) = F(B) = \overline{f}_1(B) = \int B; (\mathrm{D}\ W f \varphi_1\ x)\, \varphi_1\ \mathrm{d}x$$
$$= \int B; (\mathrm{D}\ W f \varphi\ x)\, \varphi_1\ \mathrm{d}x.$$

From 7.15 and 2.12.2 we infer the following generalization of 7.10.

7.16 THEOREM. If

$$\varphi \in \text{goodtile } W,\ \ f \in \text{tile } \varphi\ W,$$

then

$$f(\beta) \geqslant \int \beta; (\mathrm{D}\ W f \varphi\ x)\, \varphi\ \mathrm{d}x$$

whenever

$$\beta \in \text{dmn } f \cap \text{mbl } \varphi \cap \text{sb}'\ \varphi \text{ dmn } W.$$

From 7.15 and 4.21 we infer the following generalization of 7.11.

7.17 THEOREM. If

$$\varphi \in \text{goodtile } W,\ \ f \in \text{tile } \varphi\ W,$$
$$B \in \text{mbl } \varphi \cap \text{sb}'\ \varphi \text{ dmn } W,$$

then a necessary and sufficient condition that

$$B \in \text{contiled } f\ \varphi\ W$$

is that

$$f(B) = \int B; (\mathrm{D}\ W f \varphi\ x)\, \varphi\ \mathrm{d}x.$$

7.18 THEOREM. If

$$\varphi \in \text{Hull mbl } \varphi, \ u \in \text{Massable+ } \varphi,$$
$$g = \text{Gr } u \ \varphi, \ \overline{g} = \overline{\text{m}} \ g \ \varphi, \ 0 \leqslant a \leqslant b \leqslant \infty,$$
$$a \leqslant u(x) \leqslant b \text{ whenever } x \in B,$$

then

$$a \cdot \varphi(B) \leqslant \overline{g}(B) \leqslant b \cdot \varphi(B).$$

Proof. If there is some C for which

.1 $$B \subset C \in \text{dmn } g,$$

we can use 6.41.2 and 7.4 to help select B' so that

$$B \subset B' \in \text{mbl } \varphi \cap \text{dmn } g,$$
$$a \leqslant u(x) \leqslant b \text{ whenever } x \in B',$$
$$\varphi(B) = \varphi(B') \text{ and } \overline{g}(B) = g(B').$$

Then

$$\begin{aligned} a \cdot \varphi(B) &= a \cdot \varphi(B') = \textstyle\int B'; a \ \varphi \ \mathrm{d}x \leqslant \int B'; u(x) \ \varphi \ \mathrm{d}x \\ &= g(B') = \overline{g}(B) \leqslant \textstyle\int B'; b \ \varphi \ \mathrm{d}x = b \cdot \varphi(B') \\ &= b \cdot \varphi(B). \end{aligned}$$

If on the other hand there is no C for which .1 holds, then because of 6.28.1 and the remark after 6.23,

$$\varphi(B) = \infty, \ \overline{g}(B) = \infty, \ b > 0$$

and

$$0 \leqslant a \cdot \varphi(B) \leqslant \infty = \overline{g}(B) = b \cdot \varphi(B).$$

Thus the desired conclusion is obtained in either case.

7.19 THEOREM. If

$$\varphi \in \text{Hull mbl } \varphi, \quad u \in \text{Massable+ } \varphi,$$
$$g = \text{Gr } u \ \varphi, \quad \overline{g} = \overline{\mathsf{m}} \ g \ \varphi, \quad G = \text{sct } \overline{g} \ A,$$
$$\varphi_1 = \text{sct } \varphi \ A, \quad g_1 = \text{Gr } u \ \varphi_1, \quad \overline{g}_1 = \overline{\mathsf{m}} \ g_1 \ \varphi_1,$$

then

$$u \in \text{Massable+ } \varphi_1, \quad G = \overline{g}_1.$$

Proof. After using 2.10.1 to see that

.1 $$\varphi_1 \in \text{Hull mbl } \varphi_1$$

and using 2.5.5 to see that

.2 $$u \in \text{Massable+ } \varphi_1,$$

we infer our second conclusion from the

Statement. If

$$B \in \text{dmn } \varphi, \quad 1 < \lambda < \infty,$$

then

$$\overline{g}_1 (B) \leqslant \lambda \cdot G(B) \leqslant \lambda \cdot \lambda \cdot \overline{g}_1 (B).$$

Proof. Let

$$C = \mathsf{E} \ x \ (u(x) = 0), \quad D = \mathsf{E} \ x \ (u(x) = \infty),$$

and let S be such a function on ω' that

$$S(n) = \mathrm{E}\, x\, (\lambda^n \leqslant u(x) < \lambda^{n+1})$$

whenever $n \in \omega'$. Next let

$$H = \bigcup n \in \omega' \text{ sng } (S(n)),$$
$$K = \text{sng } C \cup \text{sng } D \cup H,$$
$$\psi = \text{sct } G\, B, \;\; \psi_1 = \text{sct } \overline{g}_1\, B, \;\; S = \text{rlm } \varphi.$$

Evidently

.3 $$\sigma K = S$$

and

$$K \in \text{dsjn}'' \text{ sb mbl } \varphi.$$

Because of this, 6.41.3 and 3.5.5, we learn

.4 $$K \in \text{dsjn}'' \cap \text{sb mbl } \psi \cap \text{sb mbl } \psi_1 .$$

Using .1, .2, 7.18 and the remark after 6.23, we see that

$$\psi(C) = G(B\,C) = \overline{g}\,(A\,B\,C) = 0 \cdot \varphi(A\,B\,C) = 0,$$
$$\psi_1\,(C) = \overline{g}_1\,(B\,C) = 0 \cdot \varphi_1\,(B\,C) = 0,$$
$$\psi\,(D) = G(B\,D) = \overline{g}\,(A\,B\,D) = \infty \cdot \varphi(A\,B\,D),$$
$$\psi_1\,(D) = \overline{g}_1\,(B\,D) = \infty \cdot \varphi_1\,(B\,D) = \infty \cdot \varphi(A\,B\,D).$$

Again using .1, .2 and 7.18 we see that if

$$n \in \omega' \text{ and } \beta = S(n),$$

then:

$$\lambda^n \cdot \varphi(A\,B\,\beta) \leqslant \overline{g}\,(A\,B\,\beta) = G\,(B\,\beta) = \psi\,(\beta) = \overline{g}\,(A\,B\,\beta)$$
$$\leqslant \lambda^{n+1} \cdot \varphi(A\,B\,\beta);$$
$$\lambda^n \cdot \varphi(A\,B\,\beta) = \lambda^n \cdot \varphi_1(B\,\beta) \leqslant \overline{g}_1(B\,\beta) = \psi_1(\beta)$$
$$= \overline{g}_1(B\,\beta) \leqslant \lambda^{n+1} \cdot \varphi_1(B\,\beta) = \lambda^{n+1} \cdot \varphi(A\,B\,\beta);$$
$$\lambda^n \cdot \varphi(A\,B\,\beta) \leqslant \psi\,(\beta) \leqslant \lambda^{n+1} \cdot \varphi(A\,B\,\beta);$$
$$\lambda^n \cdot \varphi(A\,B\,\beta) \leqslant \psi_1(\beta) \leqslant \lambda^{n+1} \cdot \varphi(A\,B\,\beta).$$

Accordingly, there is now no doubt that

$$\psi_1(\beta) \leqslant \lambda \cdot \psi\,(\beta) \leqslant \lambda \cdot \lambda \cdot \psi_1(\beta)$$

whenever $\beta \in K$. Because of this, .3 and .4 we conclude

$$\overline{g}_1(B) = \psi_1(S) = \Sigma\ \beta \in K\ \psi_1(\beta) \leqslant \Sigma\ \beta \in K\ (\lambda \cdot \psi\,(\beta))$$
$$= \lambda \cdot \Sigma\ \beta \in K\ \psi\,(\beta) = \lambda \cdot \psi\,(S) = \lambda \cdot G\,(B)$$
$$= \lambda \cdot \Sigma\ \beta \in K\ \psi\,(\beta) \leqslant \lambda \cdot \Sigma\ \beta \in K\ (\lambda \cdot \psi_1(\beta))$$
$$= \lambda \cdot \lambda \cdot \Sigma\ \beta \in K\ \psi_1(\beta) = \lambda \cdot \lambda \cdot \psi_1(S) = \lambda \cdot \lambda \cdot \overline{g}_1(B).$$

7.20 THEOREM. If

$$\varphi \in \text{Hull mbl}\ \varphi,\quad u \in \text{Massable+}\ \varphi,$$
$$g = \text{Gr}\ u\ \varphi \in \text{tile}\ \varphi\ W,$$
$$\varphi_1 = \text{sct}\ \varphi\ A,\quad g_1 = \text{Gr}\ u\ \varphi_1,$$

then

$$g_1 \in \text{tile}\ \varphi_1\ W.$$

Proof. Let

$$\overline{g} = \overline{\text{m}}\ g\ \varphi,\quad \overline{g}_1 = \overline{\text{m}}\ g_1\ \varphi_1,\quad G = \text{sct}\ \overline{g}\ A.$$

According to 7.19 and 5.8,

$$\overline{g_1} = G \in \text{contile } \varphi_1\ W.$$

Accordingly

.1 $$\overline{g_1} \in \text{tile } \varphi_1\ W \subset \text{pretile } \varphi_1\ W.$$

Now let

.2 $$F = \text{dmn } g_1$$

and notice that

$$\text{dmn } g \subset F.$$

Since

$$\text{Webalm } \varphi\ W \text{ dmn } g,$$

we now know

$$\text{Webalm } \varphi\ W\ F$$

and hence

.3 $$\text{Webalm } \varphi_1\ W\ F.$$

Now because of .1 and .2,

.4 $$F \subset \text{dmn } \overline{g_1} = \text{tiled } \overline{g_1}\ \varphi_1\ W,$$

and because of 6.41.1,

$$g_1 = \text{strc } \overline{g_1}\ F.$$

From .1, .3, .4, .5 and 4.14 we conclude

$$g_1 \in \text{tile } \varphi_1 \ W.$$

More general than 7.12 is

7.21 THEOREM. If

$$\varphi \in \text{goodtile } W, \ \ u \in \text{Massable+ } \varphi,$$
$$g = \text{Gr } u \ \varphi \in \text{tile } \varphi \ W,$$

then

$$\text{Al } \varphi \text{ dmn } W \ x \ (\mathsf{D} \ W \ g \ \varphi \ x = u(x)).$$

Proof. Let

$$A = \text{dmn } W, \ \ \varphi_1 = \text{sct } \varphi \ A, \ \ g_1 = \text{Gr } u \ \varphi_1,$$
$$\overline{g} = \overline{\mathsf{m}} \ g \ \varphi, \ \ \overline{g}_1 = \overline{\mathsf{m}} \ g_1 \ \varphi_1, \ \ G = \text{sct } \overline{g} \ A.$$

According to 7.19,

$$G = \overline{g}_1,$$

and according to 7.13 and 7.20,

$$\varphi_1 \in \text{solidtile } W, \ \ g_1 \in \text{tile } \varphi_1 \ W.$$

Thus according to 7.12,

$$\text{Alm } \varphi_1 \ x \ (\mathsf{D} \ W \ g_1 \ \varphi_1 \ x = u(x)).$$

From this and 5.15.1 we infer

$$\text{Alm } \varphi_1 \ x \ (\mathsf{D} \ W \ \overline{g}_1 \ \varphi_1 \ x = u(x)).$$

From this and .1 we infer

$$\text{Alm } \varphi_1 \ x \ (\mathsf{D} \ W \ G \ \varphi_1 \ x = u(x)).$$

Thus

$$\text{Al } \varphi \ A \ x \ (\mathsf{D} \ W \ G \ \varphi_1 \ x = u(x)),$$

and hence because of 5.14.3,

$$\text{Al } \varphi \ A \ x \ (\mathsf{D} \ W \ g \ \varphi \ x = u(x)).$$

8. INDEX.

References ending in 'R' refer to *Runs*[1].

Z

Greek

Signs

9. NOTES.

1. H. Kenyon and A. P. Morse, *Runs*, Pacific J. Math. *8* (1958) 811–824. References ending in 'R' refer to *Runs*.

Errata:

820R line 1, '$\ni$' should be '$\in$';
823R line 13, first 'lm' should be '$\underline{\text{lm}}$'.

2. W. W. Bledsoe and A. P. Morse, *Product measures*, Trans. Amer. Math. Soc. *79* (1955) 173–215.

3. D. C. Peterson, *Undersums*, Thesis, University of California at Berkeley, 1965; especially Theorem 6.20.

4. T. J. McMinn, *Restricted measurability*, Bull. Amer. Math. Soc. *54* (1948) 1105–1109. Theorem 3.4.

5. A. P. Morse, *Perfect blankets*, Trans. Amer. Math. Soc. *61* (1947) 418–442. Definition 2.10.

6. C. A. Hayes, *Differentiation with respect to φ-pseudostrong blankets and related problems*, Proc. Amer. Math. Soc. *3* (1952) 283–296.

7. R. de Possel, *Dérivation abstraite des fonctions d'ensemble*, Journal de Math. pures et appliquées *15* (1936) 391–409.

8. H. Hahn and A. Rosenthal, *Set functions* (University of New Mexico, 1948).

9. References ending in 'C' refer to W. W. Bledsoe and A. P. Morse, *Some aspects of covering theory*, Proc. Amer. Math. Soc. *3* (1952) 804–812.

10. A. P. Morse, *A theory of covering and differentiation*, Trans. Amer. Math. Soc. *55* (1944) 205–235.

11. J. L. Kelley, *General topology*, Van Nostrand (1955).

George Washington University
Washington, D. C. 20006

University of California
Berkeley, California 94720